BOSTON
PUBLIC
LIBRARY

DOWN FROM THE
TUSSOCK RANGES

Also by David McLeod:
The Tall Tussock (1959)
Many a Glorious Morning (1970)
Alone in a Mountain World (anthology, 1972)
Kingdom in the Hills (1974)

DOWN FROM THE TUSSOCK RANGES

David McLeod

These I remember, with the wind that blows
Forever pure down from the tussock ranges
JAMES K. BAXTER, *To My Father*

WHITCOULLS PUBLISHERS
CHRISTCHURCH SYDNEY LONDON

FIRST PUBLISHED 1980

© *1980 David McLeod*

WHITCOULLS PUBLISHERS
CHRISTCHURCH   SYDNEY   LONDON

All rights reserved. No part of this publication may be reproduced, stored in a retrieval system, or transmitted in any form or by any means, electronic, mechanical, photocopying, recording or otherwise, without prior written permission of Whitcoulls Publishers.

ISBN 0 7233 0640 0

Printed in New Zealand by Whitcoulls Ltd

To Roy the first and Roy the second
To Bruce and Jill and faithful Kay
To Lady, June and Duff and Boss
and many more of my companions
who also walked by right upon the hills
for no reward except the love of it

ACKNOWLEDGEMENTS

Acknowledgement is made to Mrs J. C. Baxter for permission to reproduce the lines from *To My Father* (from *The Fallen House*, Caxton Press) which appear on the title page. Also to Pegasus Press for permission to reproduce from Alistair Campbell's *Mine Eyes Dazzle* those lines from *Now Sleeps the Gorge* which appear on page 59.

CONTENTS

Introduction		9
1	The High Country Committee	15
2	Wartime Problems	24
3	Soil Conservation	39
4	Mock Warfare	48
5	Out on the Hill	59
6	The Winter of 1945	69
7	Cooks and their Ways	84
8	Journey into the Past	96
9	The Corridors of Power	104
10	Dawn of a New Light	113
11	The Sheep-farming Commission	128
12	Those Bloody Birds Again	138
13	Water, Water Everywhere	146
14	A Dream Come True	156
15	Bombo	163
16	Disaster—or Blessing?	174
17	Revolution	182
18	It Takes All Sorts	189
19	The Grasmere Dog Trials	202
20	New Ideas	215
21	No Blessing Here	227
22	Irrigation	238
23	Farewell	245

INTRODUCTION

No baser vision their spirit fills
Who walk by right on the naked hills.
CHARLES MORGAN

IN THE LAST thirty years or so a great deal has been written about the New Zealand high country and it might be thought that the term 'high country' might apply to almost any part of so mountainous a land. In fact, what is meant is really a type of life rather than a definable area, though it is of course confined to a rather narrow strip of land on the eastern side of the Southern Alps from Marlborough to Southland.

When the High Country Advisory Committee was formed in 1940 we sought fruitlessly for some definition which would determine what was eligible as a high-country property. Was it altitude? There were lots of areas in New Zealand of similar height but having no sort of common problems. Was it liability to snowfalls? There were many places in the area of high-country farming which had snow no more often than the Canterbury Plains. Was it the farming of merino sheep? There were flocks of many diverse breeds, though few admittedly without merino blood. Was it area? There were high-country runs from 1000 to 100,000 hectares. Was it Crown leasehold tenure? There were freehold properties and tenancies of public body endowment land as well.

There seemed no single fact which linked us all in common interest and we ended by defining a high-country run as a property on which the production of wool and store stock was the main source of income and which might be liable to losses from snow.

As it turned out, there was little real need for a definition. Those who shared our problems and concerns attended our meetings and took part in our debates; those who did not, left us to ourselves.

The high country, in fact, includes a wide range of types of run. It encompasses the rugged rock-bound peaks of the Seaward Kaikouras, the vast ranges and valleys of Molesworth and St Helens (now united in one huge cattle station), the deeply forested valleys close to the main divide (Lake Sumner, Esk Head, Lochinvar) and the great gorges of the Canterbury rivers—the Waimakariri, the Rakaia and the Rangitata. Then all of a sudden there is the dramatic change to the huge, almost-desert basin of the Mackenzie Country with its towering rampart of peaks which always dominate the scene. Even here there is a huge range of land types, from the dry warm ranges of its eastern boundary which Mackenzie traversed with his stolen flock, to the precipitous rocks of Mount Cook Station from which the shepherds and their dogs look straight across to the blue crevasses of the Tasman Glacier. Further south again, the merino farmers of Central Otago, stalking the dry moon-landscape of their rabbit-eaten hills, find a common bond with the men of the great southern lakes, Wanaka, Hawea and Wakatipu, although the land they work on is as different as if their farms were 2000 kilometres apart. Every high-country property is different; size, aspect, altitude, rainfall, soil type, nothing can be reduced to an average or norm, and perhaps the only really common factor is the type of man who lives and works there. This is the bond above all others which unites the dimensions of this far-flung territory—the bond of the hills—the love of 'country' for its own sake rather than for its possession, for many of them will never possess an acre of it, much as they would like to.

It is this dedication that kept men striving through the slump of the early thirties, it was this that sustained them through the bread-line prices of the war and it was this that made them unite to keep the station life going when it looked as if its primitive system was no longer viable in the changing world which emerged after that grim holocaust.

It is a never-ending source of wonder to me that the stations have survived the invention of synthetic fibres and emerged surprisingly unchanged as one of the strongest sectors of New Zealand farming. The high country is still unique; there is no other place in the world where the life is quite the same, where man, dog and sheep are united in a partnership to live on the very fringe of habitable land and defy the elements which threaten them at every turn. Snow and ice, fire, flood, avalanche and landslide—we meet them all.

In spite of the retirement of large areas of land classified as unsuitable for grazing, many of us can still step out as the sun's disc breaks a far horizon on the blue shingle of a mountain top and stand triumphant in our isolation while the world still sleeps beneath our feet. We can still stride across gulch and spur and shingle scree, hunting elusive sheep with cunning and faithful dogs—a far worthier and more difficult feat than seeking wild animals to kill. We can still enjoy the feeling of achievement when the long day ends with a thousand sheep stirring the sunlit dust as they race down a spur towards the open gates of a paddock. The sweat on our faces and the blisters on our feet are forgotten in the memory of work well done by dog and man.

At other times we are privileged to ride up some lovely stream-bed where there is no sound but the tumble of the water and the clink of the iron-shod hoof and look up and up to where the grey rocks above us carve pieces from the sky. In the bush country we can step silently among the black-boled pillars of the forest where the dark floor beneath our feet is dappled with shafts of sunlight sifting through the green canopy above; or step out of its semi-darkness into the blazing sunlight of some pakihi where a dozen sheep, some cattle or a group of scattered deer may surprise us by their presence. In the treeless wilderness of the Mackenzie Country the eyes are ever drawn into the distance, from the sparse tussock flats upon which we stand, to the maze of interlocking spurs whose hazy outlines are painted in delicate shades of umber and mauve and gold. The feeling of space is paramount. The sky is huge, the distances immeasurable and the sheep we seek are almost invisible in this immensity.

Many of the men who work in this environment have no skill in words to phrase their enchantment and often they do not even know the names of the trees and shrubs and herbage that they live with every day; but none of them, I think, is indifferent to his surroundings or fails to absorb with a kind of inarticulate wonder the ever-changing beauty of the unspoiled nature around him. Even today mechanical things are secondary in the work of these men; sheep and cattle, horse and dog are still their absorbing interest and the tools they work with are more often than not boot and saddle and the simple inventions of an age-old culture rather than the complicated implements of the modern world. Nonetheless they can use these things when they are required, and part of the success story of the high country since the last war has been their ability to discern how modern methods and machines can improve the old techniques without destroying the whole structure of the life.

The most profitable high-country station is still the one with the greatest simplicity of management. One where the ancient art of extensive pastoral grazing has been preserved and where concentration on the welfare of the stock has been the main consideration of all development. This book is an attempt to carry on the story of the high country through the life of one family and one property and through my own involvement in the modernisation of high-country pastoral farming, preserving it, as little changed as possible, for future generations.

The rough, musterer's life of camp and swag and packhorse team made famous by men like Peter Newton and Bruce Stronach is less evident today, but the shepherd must still stride across the hills with his team of hardy dogs and the merino and halfbred sheep are still as wild and elusive as they were when I first met them.

For the theme of this book I have chosen an extract from a verse by Charles Morgan, better known for his prose than his poetry. For me it epitomises high-country pastoral farming. Once a man is out on his mustering beat no base or shameful thoughts can cloud his mind. He has a job to do which is more than sufficient to fill it. He has no human antagonist or competitor. He is alone with the primeval world which nature, or God—for those who prefer a divine creator—has provided for the development of all forms of life. To that extent he walks by right, although in the crowded communities of the modern world he needs the sanction of our laws and customs to make use of the wilderness for his personal benefit.

I believe that association with a particular piece of land is a fundamental need for human beings, as it is for animals. Sadly there are now far too many human beings for more than a small minority of them to enjoy such an association; but it is characteristic of all primitive communities that they treasure their ties with a particular place. Nostalgia for the sights and sounds of our place of birth is seldom absent from our feelings.

> Breathes there the man, with soul so dead,
> Who never to himself hath said,
> This is my own, my native land!

Walter Scott knew the deep-seated longing for the wild places of Scotland, where primitive clans kept sacrosanct their own particular stretch of wilderness. The Maoris, torn between the delights of modern civilisation and their own sad longing for the tribal lands over which they fought, know it too. Even the feudal lords of outdated European societies had to have place names tacked on to their titles, and until recently in England this anomalous custom has been preserved even if the lord-ship should be over nothing more than a cottage in a country village.

Nowhere perhaps in the modern world is this longing so well fulfilled as in the high country of New Zealand. Nowhere can a man be so nearly monarch of all he surveys as there, and although there are many who are jealous of this privilege few of them would undertake the arduous work except for recreation, and fewer still would give up the attractions of city life in order to endure the loneliness and privations which high-country life entails.

That there are still young men with the longing in their hearts is our guarantee of the future of this life. When young men become too feeble and effeminate, then it will surely die; but after 120 years there is little sign of that. Long may it remain; and I for one am everlastingly grateful that chance or instinct led me to a land and a life that I would not have exchanged for any other.

Chapter 1

THE HIGH COUNTRY COMMITTEE

> Who hath measured the waters in the hollow of his hand, and meted out heaven with the span, and comprehended the dust of the earth in a measure, and weighed the mountains in scales, and the hills in a balance?
>
> *Isaiah*, 40, 12

WHEN C. L. Orbell and I bought the Grasmere and Cora Lynn runs in 1930 few people outside the small circle of scientists at the universities had any real conception of the history of the mountain lands of the South Island. European pioneers of one sort and another had come and explored and exclaimed about their rugged beauty or assessed them as potential farm lands, often, like Samuel Butler, appreciating both viewpoints and indulging in greater or less exaggeration. It would be safe to say that with hardly an exception they accepted what they saw as no less established and permanent than the lands they had left behind, whose human modification was of long standing. They did not realise that human modification had already taken place, because in this virtually unoccupied land there was no evidence of it nor, apparently, any need for it.

Besides this they had no knowledge of the substantial climatic changes which had taken place comparatively recently. Men of education like Butler would have known that there had been an ice age in the last few thousand years and would have recognised its marks upon the landscape which they explored, but it is only in the latter half of the twentieth century that scientific work has revealed either the extent of Polynesian burning or the climatic and vegetation changes which have taken place since these rugged ranges were freed from the rivers of ice which filled their valleys, licking out like dragons' tongues into the plains from time to time. How could they know that in the 14,000 years since the dragons retired into the farthest recesses of the alps a whole series of major vegetation changes had taken place? Nor could they have judged, even if they had known, what the current trend was when they arrived. Most of them gave no thought to such matters and proceeded to make the land habitable for themselves and their flocks; first by fire, to subdue the rampant vegetation, then by building shelter for themselves, and finally by subdividing and cultivating the land around their 'stations'.

So the life of the high-country stations began, and with its ups and downs as the price of wool and the development of the frozen meat trade raised and plunged its profits, it prospered reasonably well till the end of World War I. There were reports of deterioration from time to time, and criticism of the practice of burning to produce a 'green bite', and frantic alarm at the menace of the rabbit, but the first official action came when the Southern Pastoral Lands Commission was appointed in 1920. This commission was charged with the examination of terms and conditions of occupation as well as the study of the deterioration and depletion of the vegetation and specifically the effect and desirability of burning.

It listed the causes of deterioration as follows:
1. Burning the tussock, especially in the wrong season.
2. Overstocking with sheep.
3. Continuous grazing for seventy years without attempt at improvement.
4. Allowing rabbits to become numerous. [The most potent cause of all.]
5. The tenures under which the land is and has been held, and some of the conditions of such tenures.

The report which it prepared was concise, simple and clear both with regard to administration and management; but there is one paragraph which illustrates the commission's failure to study the available knowledge—not great admittedly—of the past history of the land:

'In primitive New Zealand the tussock grassland would

undergo no other changes than those brought about by the plants themselves under the action of the environment to which they were subject. There is no need here to go into the life history of the tussock grassland as suggested by scientific research. Suffice to say that the tussock form eventually became the plant form *par excellence*, and through that form occupying nearly all the ground there was little cover for any other species to increase its rate of occupation.' In other words the tussock grassland had attained its climax, and was fitted to endure so long as its environment remained constant.

It is surprising that even Leonard Cockayne who was a member of the commission subscribed to this statement.

The result of the commission's report was apparent in the Land Act of 1924 which embodied a number of recommendations on tenures and administration. But except for an intensification of the runholders' own attack on the rabbit problem (aided by fairly high prices for rabbit skins and the wide use of strychnine poisoning), and the generally accepted restriction of burning to a period in the spring, little change came over the management of the stations. It was still a basic tenet of the high-country management that tussock should be burnt every five years and that the more sheep that could be crowded into the high basins in summer the better for the run as a whole. The only touchstone for judging land use was the condition of the sheep in the autumn.

When I took over the management of my little 'kingdom' in May 1930 I had no other task than to perpetuate the traditional system of management, making use of every available square yard, however rugged or inaccessible, and of every plant that sheep would eat, no matter what its value might be to the environment as a whole. In this endeavour I spent the first ten years, struggling against the economic disasters of the early thirties and slowly coming to realise the part that depletion and nutrition played in that struggle.

The following thirty years passed in a slow readjustment to the realities of the situation not only in the case of my own treasured home but in the wider sphere of the high country as a whole, where our own situation was duplicated all down the

Southern Alps from Marlborough to Southland. During this time our knowledge slowly increased, not only through studies of the soil, plant and animal relationships of the present and future but through searching back into the past for the history of our country—first through the carbon-dating of material to establish the periods of fire, and later through scientific analysis of pollen deposits in swamps, to date the changes in vegetation which have taken place in the last 15,000 years.

People like simple generalisations and I am often faced with remarks like 'There has always been erosion', to which I reply, 'What do you mean by always?' Again they say, 'This country was all covered by forest.' True perhaps, but when? Certainly not when the first sheep farmers arrived, which is what they mean.

All this was to become of vital importance as pressure developed to recognise that the question of soil conservation would have to be faced in New Zealand.

Before this happened, however, the high-country men took their own initiative to deal with the problems which beset them, and in this movement I became immediately involved. The leadership involved two men, Tom Burnett of Mount Cook Station, in the very heart of the highest country in New Zealand, and Bob Todhunter, one of our most successful merino breeders, who lived at the windswept portal of the Rakaia Gorge and who also owned Upper Lake Heron Station in the Ashburton Gorge. To which of these men should go the credit I do not know but Tom Burnett, as the MP for Temuka, was in a position to approach Ministers and in April 1939 he persuaded Mr Lee Martin, Minister of Agriculture in the Labour Government of the time, to come to Tekapo and meet a gathering of runholders hastily assembled from every part of the high country.

We sat in rows on benches facing the veranda of the hotel where the Minister and his supporters and a few of our own leaders gathered to address us. A photograph taken by Donald Burnett, Tom Burnett's son, shows the gathering of serious men all dressed alike in suits and their best hats in the dusty

THE HIGH COUNTRY COMMITTEE 19

sunshine of a non-week-day, and a *Press* cutting briefly reports the Minister's sympathetic hearing of the problems presented.

There were a lot of us, considering the distance that many had to come, and the numbers reveal the widespread anxiety which we felt about the future of our livelihood.

In 1937 Molesworth Station, the largest single holding in the whole high country, had been abandoned on account of rising costs. How much the rabbits had contributed to its falling production is much disputed, but in that cataclysmic failure many of us saw the doom of other isolated runs. Where would musterers come from if these rugged training grounds were abandoned? We all had men who had cut their teeth and wrecked their boots on Molesworth's broken shingle.

Then there was the rabbit problem itself. Strychnine poison had taken the place of guns and dogs and traplines, with striking results in many cases, but some people believed that new methods were required and a more forceful policy which did not depend upon a high price for skins to make it work.

There was the spectre of falling production. Sheep numbers had fallen since the exhaustion of primeval fertility and the replacement of palatable species of grass by those that were tolerant of fire and grazing, and many thought that wool production per head was lower also, though this was often difficult to prove.

It was difficult to indicate what a helpful government could do about all these problems, but we did point out that the Department of Agriculture appeared to be far too busy studying the problems of farmers in the plains to spare a thought for the unpromising mountain sides which produced nothing more valuable than a few thousand bales of merino wool and a splendid race of men!

Where the meeting was on stronger ground was on the administrative side, and there were many present with a lot to say and little of it complimentary. The administrative body, the Lands and Survey Department, had developed out of a department whose original function lay in surveying lands of the Crown and leasing them to eager settlers anxious to get a piece of reasonable sheep country when all the lower lands were taken up. The Land Act of 1924 had been a great improvement on the previous one of 1908 but the attitude of the department was still coloured by the fact that all the Commissioners of Crown Land—the local administrators—had joined the department as surveyors. They knew nothing of farming—or so we said—and the land boards over which they presided were composed of men with ordinary farming or political backgrounds. Only a real high-country man could understand the peculiar problems of our country and there were none on any of the boards. This was something we could get our teeth into, and aided by a few who had special grievances we endeavoured to gnaw Mr Lee Martin. Politicians are tough and indigestible, however, and when the Minister rose to reply he escaped the ring of hungry teeth by saying that we had invited the wrong man to our feast and that in fact his colleague, the Minister of Lands, was the one whose blood we should be howling for. He did, however, assure us that he or someone else would be happy to come to an annual meeting where our problems could be discussed and solutions proposed.

Balked of our prey, we watched it vanish in a cloud of ministerial dust and gathered round to elect a committee which should lay the groundwork in each district for a larger and more effective meeting the following year. Next year the country was at war but nevertheless the meeting took place as planned and this time we were better prepared and what's more we had the right Minister—the Hon. Frank Langstone, Minister of Lands. As a member of the Runholders' Committee I went down for preliminary discussions, arranging for my wife, Mary, to be brought by one of our neighbours. When she arrived, rather late, she explained that they had come through the Lyndon Road which leaves the West Coast Road at Lake Lyndon and emerges about 300 metres lower down at the Lake Coleridge homestead after a rough and tortuous passage through creeks and narrow cuttings. It's different today, though even now it can be hazardous in bad weather, but in those days it was quite an adventure. Her companion had a powerful International half-ton truck—what would now be called a 'ute'—but his driving was reminiscent of handling the wool waggons, upon which he had gained his early experience, rather than of the more delicate manipulation of clutch and accelerator. At the first deepish creek he plunged in in top gear: the truck leapt wildly in the air with its wheels spinning dizzily and stopped dead when it hit the water again. With all the plugs wet with the splash, it was some time before he could get it started again; even then he had to come out backwards and give it another go.

'What a pity he didn't let you drive,' I said, because Mary was a capable and experienced driver.

'Well,' she said, 'I really thought he would be better with his hands on the wheel!'

We both laughed because his fondness for ladies' legs was well known. It was to be confirmed in the near future.

The old Tekapo Hotel, now long buried beneath the waters of the lake, had a limited number of rooms and only the ministerial party and a few of the senior runholders could be housed there. The rest of us were allotted the traditional Public Works tents of the period. They consisted of a wooden floor and wooden walls about one metre high, surmounted by a tent roof. Snug enough in fine weather but airy, to say the least, in the wintertime. One thing they were not, and that was soundproof, and as they were placed fairly close together in several ranks it was possible to overhear all that was said in the adjacent tents.

Our preliminary discussions had centred on the administrative side of our problems and they were to be illustrated by some figures submitted by Charlie Parker who owned Holbrook, not far from Tekapo, and managed Rollesby, which belonged to his wife's family. He had prepared a comparison of the rents and carrying capacities of the Mackenzie Country stations in 1896 and 1936. It was a difficult calculation because subdivision had occurred, boundaries had changed and rabbits had spread, but the burden of his song was that rents varied from 7d (6c) per sheep to four shillings (40c)—in the case of Rollesby—and that the Lands Department had no touchstone by which it judged productivity, but demanded as much as it could get and made no allowances for snow or local conditions. There was wide support for his complaint from other districts, and in the end it was resolved to ask the Minister to appoint separate high-country land boards whose members should be drawn from practical high-country men. Failing this we decided to ask for a judicial tribunal, to which we could appeal against decisions of the boards.

Other problems were discussed, notably that of reserving manpower for mustering, shearing and other active work since the drain into the armed forces had already taken its inevitable toll.

Mr Langstone, when he came, was accompanied by some of the 'top brass' of the Lands and Survey Department including the Under-Secretary for Lands, Mr Macmorran, and the commissioners for Canterbury and Otago. Naturally our attacks on their administration were not received with any enthusiasm and on their advice the Minister turned down our request for separate boards. However, in mitigation, he threw the dogs a bone which in the years that followed grew into a whole animal which, like Adam's rib, proved to have teeth and a will of its own which secured many notable achievements. He agreed that he would accept an advisory board consisting of high-country runholders with whom he and his department would discuss the problems of the area and the administration of the Lands Department. He also agreed that as vacancies became available he would appoint one high-country man to each land board from names submitted by the advisory board. There was little left to say and after the captains and the kings had departed in the usual cloud of Mackenzie dust the value of our concession was hotly debated in the bar of the hotel. There was a meeting in the evening, crowded in a small room where pipe and cigarette smoke made visibility difficult and whose main business was the selection of members of the advisory board. It was to be a smaller one than our existing runholders' committee appointed a year before, and weighted in favour of the two biggest provinces, Canterbury and Otago. When the nominations closed and the results were declared, the list read as follows:

Southland	W. J. A. MacGregor	Mt Linton
Otago	John Mackenzie	Walter Peak, Queenstown
	Willis Scaife	Glendhu, Wanaka
	Arthur Munro	Omarama
Canterbury	C. A. Parker	Rollesby, Mackenzie
	R. C. Todhunter	Blackford, Rakaia
	D. McLeod	Grasmere, Waimakariri
Marlborough	A. J. Murray	Wharenui, Clarence

I was the youngest and by far the newest of the high-country men elected and it was with a feeling of modest triumph that I went back to our tent, where Mary had retired. I found her almost hysterical with laughter, and longing to share her amusement with someone; in hushed whispers while we squeezed together in one narrow bunk she described how she had listened through the thin tent wall to her erstwhile travelling companion retailing to his tent mate the sordid details of his amorous experiences during a recent holiday in Australia. The final story was of a balmy night in a Sydney park where he lay beneath the trees with a lady of no virtue at all; a man came by, swept up the lady's handbag which lay discarded on the grass, and departed at high speed. 'And I couldn't even chase the b—,' said our friend indignantly, 'because I had my trousers down!'

So on the 19th day of April 1940 there came into being that curious informal body, the High Country Committee, which became in 1945 the High Country Committee of Federated Farmers when that organisation was incorporated. But through all its changes it still retains its original right and purpose, jealously guarded and often used—to act as an advisory body to the Minister of Lands on matters which concern the high country of the South Island.

Chapter 2

WARTIME PROBLEMS

> Work expands so as to fill the time available for its completion.
> G. NORTHCOTE PARKINSON

WITH THE OUTBREAK of war in September 1939 one fact was very obvious: there would be little labour available for the management of a high-country station. The young, fit, single men upon whom we depended for our mustering were going to be the first to be drawn into the army, and nobody was likely to make the same mistake as was made in 1914 when so many believed that the war would be over by Christmas. Nazi Germany was too well prepared to be beaten in a year or two, so we could make up our minds to be working short-handed for a long time. Fortunately for me the beginning of the war coincided almost to the day with the arrival at Grasmere of a new shepherd, a man of considerable experience, who had very good dogs and was a first class musterer. His name, Peter Newton, became a household word in later years when he began to write the books which brought the high country such widespread recognition; the first was *Wayleggo*, so named from the musterer's call to his dog to come back. I had no right to expect that I could enjoy the services of such a young and active man for long but he did in fact stay till 1942, when he could no longer resist the call to join his mates fighting in the desert.

There were many occasions during those years when Peter and I, alone or with one other man, mustered blocks which normally required five men. It took time and much walking and the skilful use of vantage points from which the scattered mobs of sheep could be spotted. Many a long run for the weary dogs was saved by the 'Nelson huntaway'—a big rock prised out of a hillside and sent bounding and crashing down to start sheep moving 300 metres below.

I remember one muster above the railway line when there were only the two of us to cover a steep face 1000 metres from top to bottom. Peter had the top and I had the bottom. I had told him not to bother with the very top as I was sure there would be no sheep up there—it was a winter muster to take the rams out and snow was lying on the top. Half way along he vanished from my sight. My 'yahoos' failed to get an answer and on some benchy outcrops of rock half way up the hill I could see a whole lot of sheep. There was nothing to do but send my heading dog for them. Luckily I had a good one, but it was a long way up a very steep face, and steering a dog becomes very difficult if it cannot see you easily. I walked out on to the railway line, where I could march right or left according to where I wished the dog to go. Unfortunately I had an audience of half a dozen railway surface men who gladly leaned upon their shovels to watch—and listen—to my efforts. My efforts were long and loud but in the end I was delighted to see Lady scramble wearily up the last bluff and make her cast round the scattered sheep. At moments like these, when the courage and devotion of a dog rewards you in a time of stress a lump comes in your throat and tears of gratitude blind your eyes. Thankfully I watched her gather the sheep together and prepare for the long descent, but at that moment a voice from my audience called out, 'Yer dog's left one behind.'

If only he could have known how pleased I was that I had got her there at all and saved myself the long and weary climb which I would have had if she had failed. One sheep was a small price to pay, and we only wanted the rams anyhow.

When Peter came down at the far end I asked him where he had got to.

'Oh,' he said, 'you were so long I walked up on to the top. I thought you said I wouldn't find any sheep up there.'

'And did you?'

'Yes, I did; about thirty ewes and two rams!'

It said a lot for the energy of our halfbred rams that they were prepared to live in the patchy snow at 1500 metres, like stags, with their little harem of ewes.

High-country management depends a great deal on utilising the natural instincts of animals and in order to foster the harem instinct in sheep I always made a practice of taking a truck-load of rams out and poking a couple here and a couple there into isolated pockets on the steeper country. A wise old sheep man told me once that he condemned the custom of hunting all the ewes on a block together from time to time when the rams were out. 'You see,' he said, 'the ewes will all climb back out to where they were before but the rams are beginning to weary after a few weeks' tupping and often stay behind.'

Another regular helper during the early years of the war was a lively old character called Bert Nosworthy. Bert was a Devonshire man and a distant relative, I think, of Sir William Nosworthy, who had been the owner of Mesopotamia when I mustered there in 1927. Bert had not prospered like his namesake, and earned his living rather precariously as a drover and a knockabout shepherd in the days when all stock were moved by road. The sight of the long dusty mobs of sheep wending their way along the wide Canterbury roads, followed by a drover in a battered gig with half a dozen skinny dogs around it was common enough, and when drought hit the province as it often does, these 'dealers' mobs' often spent weeks drifting slowly along, unhurried by the dogs, so that they could eat as much roadside grass as possible. Usually the gig contained not only the man and his swag but two or three lame or emaciated sheep as well.

Bert lived in a tiny cottage on the Main South Road near the Islington freezing works with a dear little apple-cheeked wife who must have helped him into bed many times after he had spent the evening washing the dust out of his throat at the local pub. Their cottage has now vanished in the spreading industrial suburb of Hornby.

He was very happy to take a rest from this weary work and act as paddock shepherd at shearing time or when the lambs were being weaned, and a partnership grew up between him and my wife, Mary, who used to take her old dog Pat and go and help him put sheep in the shed or take the drafting gates for him when some had got 'boxed'.

There was one memorable day when she came to help him unexpectedly as he was trying to shed some lambs. Weaned lambs are difficult and suspicious creatures because they have always followed a mother and never led the way. Suddenly wrenched from their perpetual view of the hind legs of a ewe they stubbornly refuse to walk up a race or move in any direction except back the way they came. When at last they are confronted with the dark cavern of the woolshed door it is no wonder that neither noise nor force will drive them into it.

Bert and all his dogs were struggling knee deep in terrified woolly lambs at the open door of the shed. The sun beat down on the white reflecting backs and the sweat poured down Bert's weather-beaten face. One after another he seized each lamb and flung it forward into the blackness ahead, and no sooner did it land than it turned and charged straight out again. Struggling and sweating and swearing viciously at dogs and sheep he suddenly became aware of another pair of voices: a high-pitched female one and Pat's deep-throated bark. He turned with a swear word bitten off between his only two front teeth, which stuck up and down but never met, and his shamed and chagrined face was a picture Mary has never forgotten. Naturally the language was nothing new to her, and between them they soon got the lambs going forward, one dragging a lamb in front while the other worked behind. Sheep, like many other creatures, can often be better led than driven.

Another casual who came to us for many years was Jack Gillelt, an Australian shearer and crutcher known rudely as the Great Australian Bite. I had been warned never to have him in a shearing gang as he was an inveterate grumbler and troublemaker, but on his own he was co-operation itself. He was mainly a machine man and we always shore with blades so we had no need of him for shearing, but he had one great skill with a pair of blade shears—he could shear 'stragglers' in the autumn, leaving an even covering of about 5 centimetres of wool all over the back, a very difficult thing to do. This avoided their going on through the winter to become double-fleecers at the next shearing, but protected them from the bitter cold of winter which is so dangerous for late-shorn sheep.

We still used our old water wheel to drive the machines for crutching and Jack was happy to crutch all day and fill his own pens and even to go up the water race and clean the intake if there was not sufficient water coming down. Cooks hated him because he grumbled about the tucker but he was a wonderful stand-by for us when labour was hard to get. He had one other back country place where he went regularly, Algidus—well known from Mona Anderson's books. He used to cut their firewood for the winter, and once he nearly lost his life by working in the bush alone. His axe slipped and sliced into his foot and he was reduced to crawling on hands and knees all the way to the homestead, arriving at last exhausted and weak from the loss of blood.

The other and in some ways the most important stand-by in the early days of the war was Jack Le Gros, the cook who came to us in the depths of the Depression. With children arriving at regular intervals it was absolutely necessary to have a cook for the four station hands we had and often for our own household as well, and when Jack was summoned to an army medical examination we all trembled with apprehension. If he was taken, who would cook for the men and for ourselves with far greater skill than any normal station cook? Who would light the fires and do the housework and the washing and ironing? And who, above all things, would do all this with a falsetto song upon his lips and a smile on his face? Paragons have always been scarce—another paragon like Jack was unimaginable! Mercifully for us the doctors detected weakness in his heart; he was returned to us for another year or two, and we could breathe again.

He had first come to us as station and camp cook and his reputation as a cook in the Public Works camp was his only recommendation. Really he knew little about it in those days, but he had the flair—the touch—the instinct which alone can raise a man from mere hash handling to the status of a good cook. Jack, trained by a French or Italian chef and taught to use the resources of the kitchen as they should be used, might have become a culinary artist; as it was he became a very good cook, and that title is not lightly bestowed. He could take a recipe from a book and slap the ingredients together without weighing or measuring, alter them or add to them as he thought fit and turn out something which looked and tasted as if it was meant that way. His pastry was perfect and his sponges were always light, though of course he was not frightened to use butter and eggs to make them so. I shall never forget his delight when we bought him a set of cake-icing implements. For weeks afterwards every cake he made was elaborately iced and covered with roses and fantastic designs done in every colour he could find.

In those days we had an old coal stove in our kitchen, placed in the corner opposite the sink. There was nothing near it upon which one could put anything down, so the unfortunate cook spent his life whizzing from sink to stove and stove to sink, carrying every pot and plate and dish. In between, though slightly to one side was the only table, at which the men ate, so as meal-time approached everything had to be cleared off that and thereafter the cook had to walk round the chair of the man who sat at the end. With this inconvenient arrangement Jack had to produce meals for the men, who numbered anything up to seven or eight at times, cook early breakfast for mustering, and prepare meals for our steadily increasing family, for he came just before the first child was born and left with the last. In addition to all this he did all the housework and also the washing and ironing, lit the fires and cleaned the grates, polished the furniture and cleaned the silver and had time over to take the baby for a walk.

I hardly like asking people nowadays to believe that he did all this but it's true and it happened in this country not so very long ago. The incredible part of it was the spare time he seemed to have to do things which he had no need to do, such as devise table decorations. If we had someone to stay Jack really went to town. He polished the table and cleaned the silver and rushed out into the garden to collect flowers for a vase. Unfortunately his taste in flowers was a little unconventional, shall we say? He was apt to use rather curious combinations of colours and, though we didn't like to hurt his feelings, some of them were rather hard to bear. I remember one—a flattish, round black bowl in which he had arranged concentric circles of mauve asters, orange marigolds and some pink flowers, the name of which I now forget. The mauve and gold one might have endured, but the pink was the final touch of horror!

Having arranged the table to his satisfaction, bustling in and out to add a little something or to see the fire was burning well, he would produce the most magnificent meal and then proceed to serve it, dressed in a little white mess jacket with brass buttons.

They were wonderful days! And we look back on Jack's presence with gratitude for it made the six years during which our four children were born a time of leisure and enjoyment rather than of work and worry and stress and strain as these years so often are for the less fortunate mothers of today. It couldn't last for ever, of course, and finally bad company and the usual demon brought the chapter to a close, though not without some humorous interludes.

Many people will remember the tragic story of Stanley Graham, the West Coast man who was driven by worry, war neurosis, and a persecution mania to retire to his house with a rifle and shoot at everyone who came. The siege of his house, the wounding and killing of police, and his disappearance were front-page news for days and when the police could find no trace of him they began to suppose that he had escaped into the bush and left the district altogether.

There were not many directions in which he could have gone, and someone conceived the idea that he might have gone inland, up the Arahura River and tried to get to Canterbury over the ranges. In that case he would have come down either the Rakaia via Brownings Pass, or down the Waimakariri via Brownings, Whitehorn and Harmans Passes.

WARTIME PROBLEMS

The police decided that all residents in these areas should be asked to look out for him and, like our neighbours, we were contacted by telephone. The men and I were all away and the only occupants of the station were Mary, the children and Jack. When Mary went to the telephone she was greeted by a ponderous, official voice reading to her what was obviously a prepared statement. Clearly no consideration had been given to its effect upon lonely and defenceless women. It went something like this: 'The police have reason to suspect that Graham has left the West Coast and may have travelled via the three-pass trip and down the Waimakariri. If you see an armed man or any suspicious character please inform the nearest police station.' This was not accompanied by any comforting assurance that police would be in the vicinity, waiting to grab him; it was left to the initiative of the person unfortunate enough to encounter an armed and desperate man to decide how to deal with the situation. Somewhat shaken, Mary left the telephone and went to the kitchen, where Jack was blithely rattling pots in the sink. She told him what the message was and said, 'What would you do, Jack, if Graham came to the back door?'

Jack drew himself up to his full height of 5 feet 2 inches (157 cm) and replied with these 'famous last words': 'Keep calm and treat him with contempt!'

Fortunately he was never called upon to manifest this noble and contemptuous demeanour and so perhaps it is unfair to speculate how long it would have lasted. Mary, at least, had little doubt that if there was to be any boldness it would have to come from her!

Jack had a vivid imagination and once a story had been told he could readily convince himself that it was gospel truth; gradually any embellishment of the original story became gospel truth also. One day he came back from one of his Sunday visits to Cass, rushed into the kitchen and said, 'You don't know who you've got working for you!' This remark naturally rather puzzled us and we asked him what he meant.

'Oh,' he said, 'I've found out that there's a title in my family.'

We showed some surprise, no doubt, and this made him the more determined to assert his rights to nobility. However, as he was obviously rather drunk we didn't pursue the matter and it was left to us to appreciate as best we could the honour that was being done to us.

The following Sunday he made even more rapid progress in the social scale for he came back a 'prince of the royal house of France', no less! I well rember the scene as he stood beside our dining room table, and thumped it with his fist and shouted, 'They're nothing but scum all these men here. There are only two gentlemen in the place and that's you and me.' I was deeply honoured by this tribute coming from such royal eminence and I murmured appropriate thanks. Jack went on thumping the table and saying alternately that they were scum and that he was a prince of France, so I tactfully retired— backwards, I hope!

It couldn't last much longer and the final break came after a classic scene on the drying green when he invited Mary to wash her own damned sheets, after she had remonstrated with him for dragging one of them in the mud. It was only a month before Robin was born, and when Jack flung the large wet double sheet it enveloped Mary in its clammy folds so that she looked something like a portly Roman senator who had his toga over his head by mistake. So ended nearly seven years which I think he enjoyed as much as we did, for he entered into everything that happened on the station, whether it concerned the men and their work or the family and their doings. His place was not only the kitchen—he was just as likely to be found chopping up a tree blown down by the wind or wheeling the latest baby down the front paddock in the pram, proud as any young parent.

In *Kingdom In the Hills* I told some part of the story of the Bealey Hotel. Another tragedy was yet to mark its ill-starred history, but meanwhile the new proprietor, Reg Ferguson, was a great asset to us. A first-class musterer who loved the station life, he was always ready to leave his often empty bar and come and help us muster or tail lambs or snowrake sheep when winter whitened the Burnt Face and Bealey Spur. Once the new hotel was built we often used to stay the night in it when we were mustering at Cora Lynn, instead of camping in cold discomfort in the old Cora Lynn house and cooking our tucker in a camp oven over the open fire.

The petrol rationing, which starved the Bealey of its house guests and bar patrons, had its effect on the station also. Three days a week somebody had to go to Cass to collect the mail and bread; 8 kilometres each trip, which used up a good deal of our precious ration. In the trees at the back of the house I found the remains of an old gig, which was still quite sound although it must have lain there for uncounted years. The wheels were rotten but I still had those which came off the old buckboard which Sealy Rutherford had made to fetch the mail in 1910—which shows what splendid craftsmen those old coach-builders were. I put the two together and made a smart and useful turnout pulled by my old white pony Betsy. It was as well I did because when Japan entered the war in 1941 a panic seized the Government and they impressed all the trucks which might be of use to the army. Our chief station transport was a fine new 2-ton Bedford bought in 1939, and when we reluctantly surrendered it Betsy and her cart became our chief vehicle. It was rather galling to be told by soldiers who served in Burnham Camp that rows and rows of these impressed trucks stood idle and rusting beneath the trees while their tyres and batteries slowly went flat.

Wartime improvisation took another step in 1941. We shared with our neighbour, Jim Milliken, a fearsome haymaking machine, an old stationary baler driven by a tractor with a long belt. This required an army of men to serve its ponderous activity. A horse-drawn hayrake occupied one man—usually the least intelligent—and I well remember chiding one of our wartime tractor drivers for falling behind with the raking so the whole team had to wait.

'Well,' he replied with great seriousness, 'it all depends on the horse—sometimes he goes to sleep.'

The same might have been said for the driver, I thought.

After the rake came a great wooden sweep pushed from behind by three of Jim Milliken's draught-horses, and operated preferably by someone who did not go to sleep. They delivered their load to a large pile beside the baler where there were two men, one to feed its hungry maw and one to fork the hay to the feeder. A huge, iron-shod 'horse's head' swung up and plunged down, forcing the hay down to where the ram compressed it, as in the modern machines; but the catch was that every forkful had to be timed precisely for the moment when the head was raised and the fork withdrawn before it fell. Heaven help the feeder who could not get it out in time, because the plunger had no mercy, and a twisted and splintered wreck was all that would remain of a brand new three-tined fork. This was not the only hazard of the game, however, for the bales were divided from each other by stout wooden boards which were inserted in a kind of rack by one of the men as they emerged at the back. It was the duty of the feeder to up-end this rack at the correct intervals and allow the 'horse's head' to press the board down into the empty chamber. Again the timing was critical and a fraction of a second's mistake could result in a broken board and a stoppage until it was removed. We used to find that one mistake bred others; the man would get the jitters and lose his nerve and judgement and break four or five boards one after the other.

At the back of the baler sat two men, one on either side, and their job was to push long wires between slots in the boards, one each way. The wires had a loop on one end and each man slipped the straight end through the loop on his side, one top and one bottom, and tied it with a twist. Here again there was room for error because when drifting dust obscured the view it was easy to push the wire through the slot on the wrong side of the board and then the bale would come out with the board bound to it by the wire instead of falling free. Beyond these two men there was another stacking the bales with a bale hook in each hand. Thus seven men could produce, if they worked till eight o'clock at night, about 500 bales of hay. One hour's work for two men nowadays with a modern rake and baler.

WARTIME PROBLEMS

It was not only the field staff who were employed on this exercise either. All day long a shuttle service of morning and afternoon smokos and a midday meal tore up and down between the cookshop and the paddock and when, as often happened, haymaking and shearing coincided and the shearers' cook said he was paid to cook for bloody shearers, not for bloody haymakers, the catering descended fair and square on the lady of the house no matter how many children she had.

At Christmas 1941 we had an unexpected volunteer for the haymaking, the Dean of Christchurch Cathedral, Alwyn Warren, who believed that the clergy could do more for their country in a time of crisis than to stand in a pulpit once a week and spout pious platitudes about our immortal souls. There were far too many New Zealand men fighting overseas whose souls would be immortal before long if they were not backed by the efforts of those who stayed at home.

As most people know Alwyn afterwards exchanged his pitchfork for a crozier and became the Bishop of Christchurch after serving in the Middle East with the fighting soldiers. His was a powerful and massive figure and as an addition to the seven-man team in the hayfield he was a terrific asset. Any reservations which the men may have had about a man of God working in their midst were soon dispelled when they saw him fling the bales about and join in all the jokes and badinage that any team-work soon engenders; and if the badinage included a few of the words seldom heard in church they found him undisturbed.

Not all his work at Grasmere was secular though; he took the leading part in one of the most incongruous scenes that anyone could imagine, and that was strictly a religious ceremony.

There happened to be in the railway settlement at Cass a couple of babies who had not been christened. Parsons of any denomination were fairly rare in the back country and unless the parents took the initiative and approached the vicar of the far-flung parish of Sheffield nothing much was done about any new arrivals in the world. Alwyn agreed to hold a christening service and the word was passed round that any hitherto unblessed little Christians would be welcome. It was expected that they would all be brought in their mothers' arms, or at least in a push-chair. Not a bit of it, for two stout girls of twelve and fourteen years were among the unbaptised. For many years Jim Milliken had employed a married couple, the lady keeping house for him as he and his wife had long since parted. For various reasons, including those mentioned above, the housekeeper's two daughters had never had the 'benefit of clergy' and now they appeared before the dean, who had exchanged the shirt-sleeves and patched trousers of the hayfield for the silken vestments of the cathedral.

It was not the ages of the acolytes which were so incongruous, but the surroundings. The only place in Cass large enough to accommodate the gathering was the railway goods shed, as dingy and draughty a place as all the others of its kind which grace our wayside railway stations. The congregation grouped itself on bales and boxes at one end, and at the other, with a basin on Mrs Robertshaw's best 'occasional table', Alwyn performed the ancient religious rite: first on the babies presented to him by their parents, and then on the two tall and stalwart wenches who stumped across the roughly boarded floor, their faces scarlet with embarrassment. To cap it all the godfather who undertook to oversee their spiritual welfare was none other than James Milliken himself. As the ceremony proceeded it would not have been surprising if one of the many goods trains had come thundering in and dumped a rake of waggons alongside us in the shed as they so often did.

This goods shed was the scene of a very different ceremony not long after this. Ever since the days when I began my back country career at Mount White Station across the river there had been wild cattle there. Some were really wild and dangerous but others had simply escaped a muster or two and learnt to hide in the bushy gullies at Lochinvar and had grown old and cunning in the art of dodging musterers. Jim Thompson, the manager, and his men had tried many times to corner some of these escapers because they were well bred animals and too valuable to shoot; but they had no yards or fences which could hold the cattle in, nor could they spare the time required for the task of rounding them up. At last there came along a man called Jock McArthur who accepted a contract at so much a head to get the cattle out, and it was no mean feat that he proposed to carry out. First he built a yard in the bush, stringing the wire from tree to tree in a wide circle. Then he set out to find the mobs of cattle. He had very good heading dogs and for many weeks he was content to work the cattle round, heading and driving them about until he was sure the dogs could handle them before he even attempted to put them in the yard. When at last he did yard them he had to let them out again because there was nothing there for them to eat.

So the weeks grew into months and the station hands began to laugh behind his back, saying he was just playing the fool and amusing himself for the summer; but it was he who had the laugh at the end because one fine day he appeared at Mount White with about seventy head of good cattle, having driven them all the way down the narrow 20-kilometre track from Nigger Hill. From Mount White there was another 30-kilometre drive to Cass to put them on the railway trucks to Addington market, but this was not Jock's responsibility.

The first I heard of this manoeuvre was a telephone call from Jim Thompson at Cass asking me if I would like two large steers for dog tucker. Dog tucker was always scarce and with thirty or forty dogs to feed I was reluctant to kill good wethers which we could fatten and sell for mutton. This sounded like a good offer.

'What happened to them, Jim?' I asked.

'Oh,' he said, 'they couldn't stand the drive from Mount White and they died when they got here.'

Overdriven, I thought—just typical of careless musterers!

'Righto,' I said, 'I'll be glad to have them. Where are they? I'll send the tractor and trailer over.'

The reply was rather disconcerting.

'They're in a bloody cattle truck and they're yours now so you can get them out—haw! haw! haw!'

I had been nicely caught but I wasn't going to admit to being beaten so I put a good face on it and sent Peter Newton with a gang of men and a block and tackle and the tractor, suggesting that they push the truck into the goods shed and haul the heavy beasts up to the roof with the tackle and push the truck away. Unfortunately the steers were too heavy, and the roof too low and they could not raise them high enough to push the truck away. They got them out in the end and our dogs were well fed for a week or two, but meat does not keep in midsummer and animals that die full of blood like these did keep even worse, so much of it went bad. Nobody likes to be made a fool of.

Chapter 3

SOIL CONSERVATION

> O! it is excellent
> To have a giant's strength; but it is tyrannous
> To use it like a giant.
>   SHAKESPEARE, *Measure for Measure*

THE YEAR 1941 was a fateful one. Japan entered the war—an event which overshadowed all others on this side of the world, and within New Zealand an Act was passed which has affected our high-country life ever since. This was the Soil Conservation and Rivers Control Act, and although its passing had no immediate consequences the threat which it implied to our traditional use of the high country for pastoral farming was obvious.

The first item in the list of functions laid down in the Act for the new Soil Conservation Council was 'The carrying out of surveys and investigations to ascertain the nature and extent of soil erosion in New Zealand'. I remember saying to Mary, 'They are going to find out a great deal more than is healthy for our welfare.'

For the runholders a major concern was the fact that, in order to provide finance for the administration of the Act, city communities were to be rated and, naturally enough, allowed to elect members of the catchment boards which had extensive powers to govern our lands and our lives. Who knew what impossible conditions might be forced upon us by a majority of city representatives ignorant as babes of the harsh realities of sheep farming in the mountains? Their powers were enormous; that was all we knew. Unfortunately, the authorities saw fit to awaken the public conscience to the danger of soil erosion by a campaign of propaganda directed at the unwise practices which they said had caused the dangerous state of depletion of our hill and mountain lands. Armed with horrific pictures of the man-made deserts of the American middle west they bombarded what is now known as the 'media' with abuse of the shameful methods of the New Zealand sheep farmer. He was an evil man, a deliberate despoiler of the land, and every slip and shingle slide and rocky outcrop was evidence of the havoc he had wrought. This theme, played *ad nauseam* on the great wind instruments of what amounted to a national orchestra, produced its predictable effect—a vociferous hostility towards the runholders and demands from the more outspoken critics for the complete closure of all mountain lands.

Afforestation was a favourite suggestion and the Waimakariri basin a popular place to start, because of the threat of flood waters from that river to the city of Christchurch.

The result of this campaign can be summed up in a cliché of the present day; it became what is known as 'counter-productive'. In other words it produced such resentment among the runholders that we all sought for any possible means of disproving the statements that were made; and when statements are exaggerated, loopholes are not difficult to find. Hence came the cries which we still hear today: 'There has always been erosion . . .' 'If you don't burn country from time to time there will be an accidental fire which will be much worse . . .' 'The sheep always come off that block as fat as butter so it cannot have deteriorated . . .' 'I remember that hill forty years ago and it's exactly the same.'

The Act was heavily weighted in favour of flood control, and its administration through the Ministry of Works resulted in an engineering approach to the problems, but the powers it contained and delegated to catchment boards were alarming to those who owned or leased land in the river catchments, and the threat to the high-country runholders was very real. Apart from the probability that burning would be prohibited altogether, the right which they had paid for to depasture sheep on the whole of the lands they leased from the Crown through the Lands Department could be withdrawn at the whim of a body controlled by another Government department. No wonder the Chairman of the High Country Committee, R. C. Todhunter, wrote: 'The Act would be so far-reaching and give such extreme powers in its present form that it could easily bankrupt many of our primary producers.'

Fortunately the war delayed the formation of the catchment boards until 1944, and then the common sense which is such a refreshing feature of New Zealand life prevailed in the selection of most of the members, and in particular in the appointment of the chairmen. North Canterbury elected a tough, hard-headed man, William Machin, who had known farmers all his life in Britain and New Zealand and headed a big firm of stock and station agents in Christchurch; South Canterbury chose the intelligent and ascetic Dr Randall Woodhouse who had given up a successful career in medicine to share the management of a fine hill-country property with his wife. Two more diverse types would be difficult to find, but each in his own fashion succeeded in launching an untried ship manned by untrained sailors on an uncharted sea; no mean achievement when you know the hidden rocks and swirling tides which beset their passage.

South Canterbury made a flying start by appointing a local pilot to help its Catchment Board out of harbour. This was Charlie Kerr, an ex-runholder from the Rangitata Gorge, who was deputed to act as liaison officer and explain the board's policies to the farmers whom they would affect. This went some way towards preventing them from standing on the headlands and bombarding the ship with weighty and well chosen rocks as it put to sea.

Unfortunately the formation of these boards coincided with the appointment of D. A. Campbell as publicity officer by the Soil Conservation Council. His qualifications for the job were excellent and he flung himself into the campaign to awake the national conscience with all the ardour and enthusiasm of an American salesman peddling a new product. However, his experience of soil conservation had been almost entirely in the

North Island, and the high-country men felt that he had little knowledge of the peculiar problems of their area, and that he directed too much criticism at the mistakes of the past and gave too little recognition to the changes in management and the recovery in vegetation which had taken place in the last fifty years.

I should be the last to decry his contribution to soil conservation, but the banner which he waved with such evangelical fervour did little to smooth the path of the catchment boards in getting the co-operation of the runholders.

One of the chief causes of anxiety was that the Act provided for no compensation to a landholder when an order of the board compelled him to change his land use. This power even included forcing him to change his breed of sheep, and the horror of the dedicated merino farmers of the high country at that prospect can well be imagined. The merino has something of the status of a religious sect with the appropriate prophets and high priests. Even Ganesh and Hanuman, the elephant and the monkey god, are not more revered in Hindu theology than is a champion merino ram in the high country of New Zealand. It was all very well to demand a change in land use over a small area of North Island hill country while improvements were made to stabilise it, but we were faced with the wholesale retirement of the high mountain basins which we relied on to carry our sheep in the summer, so upsetting the whole balance of a property and perhaps rendering it uneconomic. Mr Machin was more understanding of this danger than Dr Woodhouse, to whom the principle of the landholder's responsibility for the care of his land was inviolable, and when I became chairman of the High Country Committee in 1944 I had many discussions with the latter in which it was clear that our views were difficult to reconcile. I have great admiration for high principles but when one's living is at stake principles sometimes have to be modified!

The year 1944 ended with one very important event—the inspection of the upper Waimakariri basin by the newly formed North Canterbury Catchment Board. For the first time the occupiers of the six runs which had their homesteads behind Porters Pass were to meet their new masters and try to justify their stewardship of this vital river catchment.

The catchment of the Waimakariri is one of the most crucial in the South Island because of what is known as its 'down-stream values'. Situated close to its outlet into the sea is the low-lying city of Christchurch with its enormous property value threatened by any outbreak of the river from its existing course. It is now known that, not only in the period since 1850 when the city was founded, but long before, in unrecorded centuries, the river has broken out and spewed water and millions of tons of shingle over the surrounding plains. Any responsible administrators must look at the condition of the catchment of such a river and examine the reasons which led to these outbreaks.

Unfortunately, in 1944 little was known about the modifications to their environment which the early Maori settlers had made and so the whole attention of the board was concentrated on the actions of the pioneer sheep farmers and their successors. It had been recorded by L. G. D. Acland in *Early Canterbury Runs* that Joseph Pearson had burnt the country he explored in 1857 and that the smoke of his fires was visible in Christchurch. This was not surprising of course; nor'-west winds will even carry smoke from Australian bush fires and leave a smudge of grey ash on the pristine beauty of our snow fields. These and similar fires may be regarded as inevitable; even if the country had not been used for sheep farming its highly inflammable vegetation would have been burnt in some accidental holocaust during the years which followed. The question confronting the Catchment Board was not the initial fires but the practices of burning and grazing which had developed subsequently. On the face of it the record was not good. Forested slopes on the Craigieburn Range, in the Esk Valley, and at Cora Lynn, had been fired to produce extra grazing, as a host of blackened stumps bore witness. Dense patches of snow grass had been burnt in high summer by some exasperated musterer after vainly trying to find sheep amongst it—like my own man Arthur Booth the first season I was at Grasmere. Regular burning to get a 'green bite' had bared the sunny faces until the creeping shingle pushed down by the elements and the persistent feet of sheep began to obliterate the tussock, which alone could hold the soil in position. In addition to these ill treatments the virgin fertility of the land had led the early runholders to graze the maximum number of sheep which their country could support and, in order to do so they hunted as many as possible of their active merinos out on to the high and sparsely vegetated tops during the hot dry summer period in order to save their winter grazing. In this they were not much more guilty than their overlords of the Lands and Survey Department who placed no restriction on the number of sheep which could be grazed and made no attempt to prevent a runholder who wished to sell from crowding a ridiculous number of sheep on his lease in order to get the best price for the property.

It was impossible to refute these criticisms where they were based on obvious and well documented evidence but there were some statements—such as that by one 'scientist'—that the level of the vegetation on the Craigieburn Range had come down 600 metres since the settlement of that area—which could be shown by photographs to be absurd. In any case there was an unlimited field for argument as to what effect sheep-farming practices had had upon the Waimakariri River.

It is always difficult when you are looking for some specific feature not to let your eye be caught by every manifestation of it and I am sure that the Catchment Board party which inspected the upper Waimakariri could not have failed to notice every shingle slide however old, and every yellow streak which slashed the tussock slopes revealing eroding subsoil on the sunny slopes. The runholders who attended their progress did not fail to point out the reverse side of the coin where, on the shady faces, the vegetation grew strong, and alpine shrubs and young mountain beech were already reclothing the slopes once ravaged by fire.

The most satisfactory result of this expedition was that we met the men we feared and found little hostility among them.

Men like George Jobberns, who was Chairman of the Soil Conservation Committee—a kindly man if ever there was one; Lance McCaskill whose knowledge of alpine botany was almost intimidating; R. M. D. (Peter) Johnson, a runholder himself and master of a huge slice of the Torlesse Range; and Tom Preston who, as Commissioner of Crown Lands, had a foot in both camps. These were a few of the men in whose hands our future lay, and by the end of the day we began to breathe more freely and feel that our fears had been unjustified; there wasn't an ogre in the lot!

The High Country Committee had several meetings with the combined North and South Canterbury boards and gradually a policy emerged to which we could find small objection.

The boards' first announcement—that the runholders' continuing presence was necessary for the country—removed one nagging anxiety, the fear of expulsion. Then, instead of initiating the complete ban with which we had been threatened, they decided on a policy of issuing permits to burn where it could be shown to be necessary. Finally, they resolved to set up a series of trials in different areas in an attempt to determine the facts about the vegetation before adopting any specific policy. The trials took the form of fenced plots and line transects from which regular readings were taken to indicate whether the vegetation was increasing or decreasing. Not unnaturally the fenced plots mostly showed increased cover even if this was only in the form of hay. The line transects were mostly set out on depleted and sunny slopes and showed deterioration, especially above 1000 metres. I couldn't complain about these findings because I had already predicted them!

In spite of the boards' reassuring attitude we still felt the need of some protection. Their authority was enormous and although they assured us that they would never use it dictatorially they couldn't deny that they had the power to make us bankrupt. As a result we harked back to the demands we had made at Tekapo and asked the boards if they would consent to a judicial tribunal which would give us an independent court of appeal against decisions which we thought unfair.

Not surprisingly they refused to accept a proposal which reduced their authority except in so far as it concerned our financial position. Here at least we got acceptance of the idea that some such tribunal might grant us compensation.

All this is ancient history because the principle of grants and subsidies to compensate for the loss of grazing on eroded lands has long formed part of official soil conservation policy, and on the whole it is fairly and generously applied. However, this concession was not achieved without a struggle, and how often at the present time do we see similar demands for an authority to whom appeals may be made against the rulings of the bureaucrat? The creation of the ombudsman is the official acknowledgement of the individual's need for protection.

I can say that the catchment boards have never to my knowledge used their powers unreasonably; had the tribunal we requested been set up it would have had little work to do.

In June 1944 Mr Todhunter retired from the chairmanship of the fledgling High Country Committee and I took his place. There were many problems to face. One of the most important of these was the question of how to make ends meet with the very low wool prices fixed by the 'commandeer'.

There was virtually no premium for the finer wools; the top price received for Grasmere wool during the first three years of the war was only 15 pence a pound, and the average price 11 pence. With the help of the Canterbury Sheepowners Union we managed to persuade the powers that governed the 'bareme'—the price list of the hundreds of different types of wool—that an infinitesimal minimum stolen from the crossbred growers of the Dominion who made up eighty per cent of the clip, added to the prices paid to the small minority of fine-wool growers, would help to reduce what we considered a very unfair allocation of the total funds. This deed was done in dire and deadly secrecy and so the squeal of rage from the crossbred growers which might have greeted its publication was never heard. The High Country Committee had chalked up its first major success. As far as Grasmere was concerned this change resulted in a rise of about twopence a pound and added £500 or £600 ($1000-$1200) to our precarious income.

Another of the committee's main preoccupations was the menace of the rabbit. With most able-bodied countrymen away at the war the labour-intensive job of rabbiting became almost impossible and the catchment boards were well aware that they could never cure depletion and erosion as long as large areas of the country were overrun with this insidious pest. It was not until the Killer Rabbit Boards were formed and the trade in skin and carcase abolished that any progress was made and that was against the bitter opposition of a lot of people—runholders among them—who had had a very satisfactory income from the vermin. Vested interests die hard!

I used to hate presiding over meetings where some of the old diehards from Central Otago fought bitterly to get us to oppose the formation of the boards. It was difficult for me because I lived in a district where rabbits had never been a major pest. Scattered patches have appeared from time to time but seem to die out again with little more than a token attempt to poison them. Eventually we were forced to form a board, which cost twenty or thirty times what we used to spend, and we still have the sporadic patches of infestation!

It was the aeroplane and the deadly 1080 poison after the war that finally reduced the rabbit population to manageable proportions, and in the dry areas of the Mackenzie Country and Central Otago this did more to renovate the land than anything the catchment boards could do.

Chapter 4

MOCK WARFARE

> There is rock to the left and rock to the right and low lean thorn between
> And ye may hear a breech-bolt snick where never a man is seen.
>
> RUDYARD KIPLING, *Ballad of East and West*

AFTER THE ENTRY of Japan into the war the New Zealand Government recognised the danger of the whole Pacific being overrun. New Zealand itself would not have been an important conquest except that a naval base in, say Auckland, would have enabled the Japanese to dominate the whole South Pacific and cut off Australia from the United States. In addition they could have stopped entirely the flow of meat, wool and dairy products to the United Kingdom. The history of the defence of the South Pacific by the American navy is well known; but, as in Britain after Dunkirk, there were few resources in New Zealand to defend it against raiding forces or attempts at occupation. So there came into being the New Zealand counterpart of the British Home Guard, commanded by survivors from the war of 1914-18 and organised and exercised along the lines of the army drill of that far-off period. There was much humour to be got from the antics of the dug-outs and their efforts to make an effective and disciplined force out of the mixed selection of town and country men left over after most of the fit young men had departed for the Middle East and the Pacific Islands. There was also a little known and picturesque branch of the Home Guard which found enthusiastic volunteers among the back-country men of both islands. These were the Guide

Platoons, whose duties were to be the dream of every adult boy scout.

In the event of a Japanese landing they were to work from bases in the inhospitable mountains, guiding defending troops by familiar paths and serving as spies and saboteurs. I forget how many of these platoons there were in the South Island—sixteen or seventeen, I think, strategically placed in isolated areas. The upper Waimakariri region was an obvious choice of location for one platoon, with a main road and the only railway crossing the island. Visions of T. E. Lawrence and his attacks upon the Turkish railway line during the 1914-18 war immediately come to mind.

I was selected for the command of a platoon in this home area and so began a sort of weekend second life which took me into an imaginary cloak and dagger world and encouraged me to explore parts of the country where I had never been before.

There were limestone caves at Castle Hill and the old abandoned coal pits at Avoca in the Broken River Valley which might serve as hide-outs. (Not the coal pits, we decided, because they were most unsafe and usually full of water!) We even constructed a hide-out of our own, making a very damp bush hut roofed with branches and an old tarpaulin on a spur from which a look-out could be kept over the road and railway for many miles. Behind it was a palisade of grey-faced rock which we hoped would camouflage ascending smoke—a reminiscence from a boyhood book about Indian battles in the Wild West.

The Guide Platoons were first designed to consist of only ten men, if my memory serves me right, but even this number was difficult to collect from the five sheep stations in the gorge with the addition of one or two rather reluctant railway surface-men, and when it was decided later to expand each platoon to twenty-seven or so we were compelled to amalgamate with the north side of the Rakaia Gorge and move our headquarters to Highpeak Station on the other side of the range. Here we made our armoury in the new woolshed which was the pride of the manager, Ted Relph. He was fanatical about the cleanliness of his precious woolroom, and heaven help anyone who spilt a drop of rifle oil upon it.

Our next door neighbours to the south were at Mount Possession in the Ashburton Gorge and their Commander, Sam Chaffey, outdid us all with his spectacular establishment. He was an extrovert in everything he did and the Guide Platoon reflected his energetic personality.

First of all his men were mounted on horses to achieve the maximum mobility, and one couldn't escape the feeling that the Wild West novels which litter most back-country huts had some bearing on this decision. Where speed is important horses may be valuable but they are very hard to hide if secrecy is needed, and they leave indelible evidence of their passage in hoof-marks and droppings. My own feeling was that men on foot would be more effective for guerrilla warfare and a great deal easier to conceal. If they were to serve as guides they would be guiding men on foot anyway.

To add a touch of glamour and some uniformity to his mounted heroes Sam designed a wideawake hat like that worn by the Australians in World War I.

What were we trained to do? At first our instructions were rather vague and our training very sketchy but in May 1942 a training course was held on Dunedin's Wingatui Racecourse, then occupied by the Otago Regiment. Here the commanders, uniformed for the occasion as honorary second lieutenants, were put through a course of instruction by two or three tough young sergeants and a captain who had come back from the Middle East. They were a cheerful, energetic and very bawdy bunch who taught us the use of a wide range of explosives, introduced us to the newly developed tommy-guns and other automatic weapons, and initiated us into the mysteries of signals, map reading, astral path-finding, and assorted methods of murder made legitimate in war, some of them too revolting to describe. We made hand grenades and learnt to hold them in our hands after lighting the fuse and pause for the appropriate time before throwing them at a target—a difficult lesson to absorb. I remember that the first one I threw left my hand the instant I saw the sparks fizzing out of the end of the alarmingly short length of fuse—much to the disgust of my instructor.

MOCK WARFARE 51

There was a standard warning issued to us all, which consisted of a little account of the progress of men in the handling of explosives. The first stage involves feelings of considerable apprehension, nervously obeying every safety rule; one then becomes fairly confident but still obeys the rules designed for one's protection. The third stage is that of careless familiarity, in which one becomes thoroughly overconfident and reckless; and the fourth stage of being a careful and reliable expert is only reached if one survives the third! In due course I learnt the truth contained in this warning.

We ended our training with an expedition into the hilly and bushy country which comes so close to the built-up city of Dunedin. Here after various exercises and night manoeuvres we camped in an uncomfortable hut, and in the morning it became very clear that the majority of us had reached the third stage in our progress towards expertise with explosives. We were woken by some of the more lively members of our platoon, who had perhaps spent a cold and sleepless night; their morning call consisted of half plugs of gelignite fused and detonated against the back wall of the hut. An attack by the Japs could hardly have turned us out of bed quicker. This was followed by the lighting of breakfast fires with damp wood—encouraged by the addition of small pieces of gelignite, which burns extremely well and, strange to say, does not explode under those conditions. It is curious stuff, which can be detonated by a rifle bullet, particularly when it is cold, and this characteristic had been demonstrated to us already when we were learning to make road-blocks by felling trees across the road. A plaster of gelignite across the trunk, detonated with a rifle bullet, will bring a sizeable tree to the ground much quicker than a man with an axe.

The training camp ended with a night on the town in Dunedin—a city not noted for the extent and variety of its night life—and a return to our home platoons to pass on the information we had been given. One of the first developments after this was the issue of three whole tommy-guns to each platoon, together with one box of fifty cartridges, accompanied by the instruction that this live ammunition was not to be used except in an emergency. Anyone familiar with a tommy-gun will know that one good burst from each gun would have used the lot, so this niggardly issue was of no practical use even had the emergency occurred. Under the circumstances, and knowing the extreme difficulty of hitting anything with these guns until you became accustomed to them, I decided that it was far better to give the men a few practice shots so that they might have a chance of hitting a target when more ammunition became available.

This was during the period when we had only ten men based at Cass, and for the purpose of this exercise I took them down to Castle Hill, where the large limestone rocks provided a target which would show the marks of a bullet. The tommy-gun, with its short-rifled barrel, climbs upward and to the right, and the first time you use it you are very unlikely to get more than one shot on even a large target—the rest of even a very short burst flies harmlessly into the sky. I found a large, flat-faced rock and placed my men about ten metres from it. Each in turn was to fire three single shots—we could not afford a burst.

Two or three men fired, with very limited success, even though the target was 3 metres square; but some shots undoubtedly hit for you could see the dent in the rock when we inspected it after each man. Suddenly someone at the firing point remarked, 'Something went past my ear!'

'Nonsense!' I said, 'You must be dreaming.'

Then another man confirmed that he also heard a 'kind of whizz'. We looked at each other in alarm and then began to search the ground around us. To my horror we discovered half a dozen well flattened bullets, some of which had gone well beyond us. They had obviously bounced back off the solid rock, and any one of them could have hit and injured one of us. It was a possibility I had not considered and I was happy to call off the rest of the exercise and thank my lucky stars that nobody was hurt.

A more amusing but probably equally dangerous event took place in the middle of winter when snow lay on the ground for several weeks in 1943. I happened to be getting the mail from Cass one day, and as usual after collecting the mailbag from our postmistress, Robbie, and the bread and other parcels from the 'dog box', as we all called the waiting room at the railway station, I went to look at a rake of trucks drawn up on the siding, to see if any were consigned to us. It was an old complaint which we never got redressed that the goods shed in Christchurch would never specify the different consignees when they sent a 'miscellaneous' truck. All they would put on the ticket was 'various', leaving us all in turn to undo the tarpaulin covers, let down a heavy iron door and climb inside into a dirty, pitch dark cavern where it was impossible to read the names on the packages anyway. This was annoying enough for men but quite intolerable for any unfortunate woman who went to fetch the mail.

In this instance the usual problem did not arise because the first truck which I looked at was literally plastered with warning notices: 'Danger!' 'Keep fires away!' 'Explosives!' 'Handle With Care!' On looking at the ticket clipped at the end of the truck I saw: '*Consignor:* S.M.D. Christchurch (Southern Military District)—*Consignee:* Armament Supply Officer. Cass.' Who on earth could that be? Perhaps the army had a big manoeuvre planned. Perhaps they proposed to make a secret dump of ammunition in the mountains. Perhaps a dozen unlikely possibilities. Anyway there the lethal cargo stood, threatening the neighbourhood with its brooding presence and causing acute anxiety to Sam Robertshaw who was in charge of the railway yard. It stood there for a fortnight, shunned by all who came and went, and no light was shed upon its real purpose and destination until one day the bombshell fell. I got an urgent telephone call from S.M.D. to say, 'For God's sake empty that wretched waggon. New Zealand Railways are going mad about the thing being left on their precious property.'

'But why me?' I cried. 'It's consigned to the Armaments Supply Officer.'

'Well,' said S.M.D. sharply, 'you are the Armaments Supply Officer!'

'First I've heard of it,' I said. 'What am I to do with the stuff?'

'Get it off the Cass Railway Station. The Railways are going mad about it.'

'Where am I to put it?' I said plaintively. 'My armoury is over the Rakaia Gorge now.'

'We don't care where you put it,' said S.M.D. rudely, and slammed down the receiver.

Reflecting on the possibilities I came to one very clear conclusion. If the Railways were going mad about the stuff on their property it was nothing to what Ted Relph would do if I suggested filling his woolshed with high explosives. I didn't even know what the truck contained so I thought the first thing to do was go and find out. Taking a tractor and trailer I drove into Cass and, with the help of a very reluctant shepherd, I unsheeted the waggon and very carefully unloaded fourteen assorted cases. Most of them were labelled with military hieroglyphics which I didn't understand but there was one large iron case a little less than one metre square whose designation was clear; it was filled with electric detonators: about as dangerous a cargo as one could well imagine. Driving very slowly and avoiding all bumps and potholes we took our lethal cargo home to Grasmere and unloaded it into the chaff house at the stable.

After all, it had bumped up the railway line in an unsprung waggon, so there was probably little real danger—unless the stable caught fire. Curiosity prompted me to unpack the cases and I found to my considerable puzzlement that they seemed to consist mainly of 'primers'. After my training course I knew enough to recognise them. A primer is a small cartridge of very high explosive into which a detonator can be fitted; the cartridge is then inserted in a large mine containing less volatile explosive. I could not conceive that the New Zealand forces had enough of this sort of stuff to spare fourteen cases of it to sixteen Guide Platoons on the remote chance of its being required. I suspected that it was naval stuff anyway, intended perhaps for mining harbour approaches or some such thing. It had come direct from a ship in Lyttelton, and it was obvious that the gremlins had got among the consignment papers and sent it all to the wrong place.

Rather anxiously I left it where it was and requested S.M.D. to get it off my property before I went as mad as the Railways. It took two months to sort out the mess and we all breathed a sigh of relief when I was told to rail it back to Lyttelton.

Our platoon had one final fling before the Japanese threat was obviously past; one which was not without its amusing nor its dangerous moments. There was to be a mock battle between the Home Guard battalions on the north and south side of the Rakaia River, taking the form of an attack on the Rakaia Gorge bridge by the southern forces and its defence by the Sheffield-Darfield Battalion. This was to be a night exercise and we were invited to take part in it. Our commanding officer welcomed us when we arrived and said he was looking forward to seeing us in action. It seemed pretty obvious to me that there was going to be very little action we could take; it was not the sort of thing we had any special training for and we were too few to act as an effective screen for scouts on such a wide front. If we had been on the attacking side our training as a unit and our methods of infiltration and destruction might have been appropriate. However we made our way up the terraces above the bridge in the hope of being in the right place when the enemy forces came.

We lay there for a long time, seeing nothing in the pitch dark night and occasionally hearing the twanging of a fence wire, which might have been caused by men or by the numerous cattle which grazed on Bob Wightman's paddocks. At last I caught the sound of low voices, and against the faint lightness of the sky I saw two heads moving across our front. We had prepared ourselves with a small stock of our favourite weapon, the half plug of gelignite with detonator and short fuse, which we carried stuffed inside our shirts to keep it warm. I gave the word and we lit two or three of these and flung them in the direction of the voices. The bangs were gratifyingly loud and the flash of the explosives lit up two lonely figures—walking without any attempt at concealment, unarmed and distinguished by white armbands. They were obviously umpires and we decided unanimously to retire to some more salubrious spot!

During the rest of the night we saw and heard nothing except a mob of calves sheltering under a plantation. The temptation was too great to be resisted and half a dozen of our home-made grenades created a very satisfactory stampede which would have done credit to any Indian raiding party. Eventually we returned towards the bridge to find out what was happening and were greeted with the news that the enemy had sent a party to outflank our troops, sneaking round above the river, dropping neatly into our headquarters near the bridge and capturing the lot! It was just the sort of raid we should have been trying to carry out ourselves. We were invited to spend the rest of the night as best we could, and selected a stack of hay-bales in a plantation for a few hours' rest. This field day was attended by the officer in charge of all the Guide Platoons in the South Island, Dooley Coxhead, whose job it was to supervise our training and our equipment with the aid of a sergeant. That he hadn't been with us during the night goes without saying or the incidents of the umpires and calves would probably have never occurred, but he joined us in our alfresco bedroom on the hay bales. It was a draughty caravanserai and the housemaids had made the beds extremely badly, with the result that I woke as soon as it was light and began to think about at least a hot cup of tea, if not exactly a three-course breakfast.

My first job was to light a fire to get a billy going, and I used the now familiar technique of starting it with a few shavings off a plug of gelignite. This gave me a mischievous idea so I made up a plug with fuse and detonator, intending to rouse our commanding officer with an appropriate salute. In doing so I broke one of the rules for handling explosives—I did not freshly cut the end of the fuse which I was going to light. When I applied the match to it, it did not catch because it had got damp. Thinking matches were a waste of time when I had the billy fire at hand I thrust the end of the fuse into the flame for a moment. Still it did not start to fizz, so without undue haste I began to do what I should have done first—take out my knife and cut the fuse afresh. No sooner did I make the cut than I realised with horror that instead of igniting at the damp end the fuse was burning in the middle. I flung it from me in a panic and it landed not far from Dooley's rustic couch, exploding as it touched the ground. Dooley and everyone else woke up, abusing me not unnaturally for disturbing their sleep so rudely; but they didn't see my shaking hands or notice my shocked white face in the pallor of the dawn. And so I graduated from the third into the fourth stage of handling explosives—I had been just one second short of losing all my fingers.

There was an epilogue to this semi-military interlude in the quiet life of the high country and it is worth recounting because it shows the shocking power of the forces we were light-heartedly playing with.

When the Guide Platoons were disbanded and the weapons of war—suitable or otherwise—returned to our paternal S.M.D. I was left with a box of rather ancient gelignite which had been through a number of temperature changes—some of them extreme, as is common in the mountain seasons. I decided that this was liable to be what the experts call 'sticky' and might be highly dangerous. The water race which has served the station for about a hundred years—'the water race that runs uphill', as it was called—had been disrupted many times by the Cass River from which it came, and at the time I speak of there was a very large boulder lodged firmly in the intake which made it very difficult to get the water into it. This was a lump of rounded rock, water worn by the tumbling stream probably for several hundred years. It would have weighed perhaps two tonnes and measured one metre across. If I could pack the remains of the sticky gelignite firmly enough beneath it the blast should lift it out and fling it out into the main river bed, I thought.

Unfortunately there was water all round it, which I could not divert; however, with the help of my reluctant shepherd, Ernie Percy, who had little stomach for this job, I submerged the charge of eleven plugs, fitted with a detonator and about 60 centimetres of fuse, beneath the water and the rock and we shovelled as much shingle as we could on top of it. Before lighting the fuse we searched around for a safe place to shelter when the blast went off, and chose a clump of mountain beech trees a hundred metres or more away.

Crouched behind their stout trunks we waited nervously for the bang. After what seemed an age, during which I was quite sure the fuse had gone out under the water, there came the most stupendous bang, which rolled and reverberated to and fro across the narrow valley. It was followed by a shower of small shrapnel, which spattered the wide river bed from side to side and even penetrated the leafy canopy above our heads. When all was silent we emerged, expecting to see the rock lying some yards out on the shingle, but there was not a sign of it, nor could we find a fragment as evidence of its existence; the blast had disintegrated it into its primeval components as if it had never been. All that was left was a deep hole in the mouth of the water race rapidly filling as the water seeped back into it.

Chapter 5

OUT ON THE HILL

> Now sleeps the gorge, the pale moon's steaming disk
> Desolate and glimmering through the gusty mist;
> The storm that through the wind-cropped tussock
> Screams, and screams where the great hawks rest
> Upon comfortless stone their arrogant hearts.
> ALISTAIR CAMPBELL, *Now Sleeps the Gorge*

MUSTERING IN WARTIME presented many problems because of the shortage of men; these problems were compounded because that decade brought some of the worst weather I experienced during the forty years in which I managed the station. It began with the snow of 1939 and continued almost without relief until 1948. It was not only the bad winter snows like those of 1945-7 which drove me to despair, but also the storms at lambing time; all through the high country there was a shortage of sheep as a result, and stations which normally bred all the lambs they needed had to buy sheep to keep their numbers up.

In 1942 the autumn weather was appalling and we waited many days for the incessant nor'-west rain to cease before going up to our Top Hut to get the sheep off 'Powers Country'—our summer grazing for the wether flock. Normally the first day of April is the time for this expedition, for the basins are very high and the sheep must all come round and down a narrow ridge which can be blocked by snow very easily. The valleys all drop into precipitous gorges where rocks and waterfalls and dense bush make access impossible.

The day we actually went up, with our packhorse team carrying the swags and stores, the weather turned wet again before we got there. During the night it began to snow and when we awoke in the cold silent hut next morning we found our bedding dusted with fine white powder which had blown up under the roofing iron during the night. All day it continued to snow and as we had had no time to stock the hut with dry firewood—if any such existed after all the rain—we preferred to spend the day rolled in our blankets and sleeping bags and save the little wood we had for cooking. That night the snow stopped falling, but in the morning it lay 30 centimetres deep all round the hut.

We held a council of war. It was almost insane to try to muster in such conditions and if there were 30 centimetres at 1500 metres where we were, there might well be twice as much at the Bruce Saddle where we had to go. Against that, none of the men could stay for long except Reg Ferguson and my own man Peter Newton, and if I delayed the muster it might have to be done with three men instead of five. We finally decided to go and look, in the hope that there might not be many sheep in the valley of the Bruce. It was a weary expedition; down through the sopping bush trying to follow the deer track hidden by the snow which had seeped through the canopy above; wading the swollen and icy creek to get on to the Blind Spur; and climbing through the ever-deepening snow on the other side to come out on to the open top where it came half way up to our knees. Luckily it was dry powder snow and not too hard to flounder through, but by the time we got to the highest point where we could look across to the Bruce Saddle, it was a full 45 centimetres deep. There we crouched and searched the country with our field glasses, looking for the little groups of wethers we expected to see, grey against the Persil whiteness of the snow. To our relief we could see none and we could only assume that the wise old wethers had turned before the storm and made for home before it got too deep. There was not much more that we could do—at least we knew that most of the sheep would be on our next day's mustering country. The sun had come out and a light nor'-west wind began to thaw the snow and make it settle to a lesser depth, but we were a leg-weary gang by the time we got back to the hut, especially one big heavy man, Frank Kerr, who had come up from 'down country' to help us and who had not walked on the high tops for many months.

Next day a good deal of snow had melted and we were able to carry on, though the walking was hard; at least the sheep could move through the thawing snow, which packed readily beneath their feet. The third day's muster through the Jordan basins was completed without the weather breaking again and it was with great relief that I saw the long string of sheep come out on to the Top Flats, safe from the death trap in the basins above the bush. There was no time to lose because I could only have the extra men another couple of days and we still had the Cora Lynn Tops to do. One man in fact went home at once and four of us did the first day on that cold and inhospitable ridge where remnants of the previous week's snow made it difficult to get the sheep to cross the heavy drifts which formed a cornice beneath the top.

After that we were down to three men for the rest of the season until the rams went out in May. What I should have done without Reg Ferguson, living at the Bealey Hotel and always ready for a day on the hill, I can't imagine. If there were sheep to snow-rake on the Burnt Face, or keas were killing trapped wethers in the snow, or the sheep needed hunting out on Powers Country when they first went up, he would be there with his rifle ever ready in case he saw a deer—and no man saw more deer than he did, for he had eyes like the proverbial hawk.

After the winter of 1943 there were a number of avalanches in the spring. There must have been a lot of snow high up, although the winter was not a bad one on the whole. When the warmer nor'-west winds came the sodden mass which clothes the rocks on Misery came thundering down the Snow-slide Basin across the Cass, and further up the river another big mass tore a swathe through the dense bush when it swept down from a hanging valley on the Cora Lynn Tops. When we walked up that branch of the river in the autumn we saw not only a tangled heap of timber all down the slope on our right but, on the other side, a rarely seen phenomenon—the blast of air driven in front of the snow had felled the trees *uphill* to a height of 100 metres or more.

After the Powers Country muster that year Reg Ferguson and I decided to go back up the Bruce Creek because we had seen forty or fifty sheep on the western ridge above the creek. They had been too far along to get a dog up to them and moreover we were short-handed, so willy nilly we left them where they were. There were two ways to get them out—either right round through the basins to the Top Hut or down the ridge they were on, to Mid-Hill behind the Burnt Face. This was much shorter because, once on the Burnt Face, they were comparatively safe and could be left to winter there. The other way meant a long drive and probably a night at the Top Hut, for which we would have to carry tucker and sleeping bags. It is a long, long walk up to the headwaters of the Bruce and much of the river has to be by-passed by climbing round high bluffs or by some precarious deer track which skirts a waterfall or a deep pool. Elsewhere the stream is swift and icy cold at that time of year and has to be crossed repeatedly. Half way up we found another avalanche tip in which there was still a pile of dingy grey winter snow consolidated to an almost rock-like hardness. It was now late April and the snow must have blocked the river completely when it first swept down in September, forming a subterranean cavern through which the water passed for many months.

The walls of this canyon, where even the deer find it difficult to penetrate, are rich in alpine plants—treasures rarely seen where animals can reach them. Gentians and edelweiss abound and the delicate anisotome that sheep feed on to extinction. The wet rocks beside the waterfalls, where we had to pass our dogs up hand to hand, are covered with green and golden moss and cascades of white flowering *Helichrysum*. There are masses of white and scarlet snowberries that the keas love and in the open spaces great carpets of *Celmisia* of several different kinds. When we followed deer tracks through the bush we found the dark tree trunks dappled with silver and green and golden lichens and the grey streamers of the old man's beard swung from their gnarled and twisted branches. Sometimes the sunlight reached down into the depths and lit up the bright colours like a stained-glass window and in other places it was almost too dark to pick a track among the trees. As we got further up, the valley got wider and we walked through open bush where flocks of small birds accompanied us, twittering excitedly through the upper branches where the sunlight picked out myriad flashes of green wings when they swirled like wind-blown leaves. We had little time to pause and enjoy all this unspoilt loveliness but when we reached the Junction Creek, which was where we must leave the valley and start to climb, we allowed ourselves a rest and a boil. We had been walking nearly four hours, and the day's work still lay ahead.

Above the junction on the Mid-Hill side is a giant shingle scree which must be 1000 metres high from the ridge to where we stood. Up this we had to climb until we were above the bushline and then try to drive the sheep, which were scattered through rocky bluffs, along above the bush. It did not look too difficult and from where we were the sheep were plain to see. It was not long before we discovered how wrong we were! Sheep, like all other animals, keep to their well-marked tracks. Their small, agile feet level a path across running shingle slides, and where there are rocky bluffs they will only enter them when they know that pioneers have passed before and their eyes and noses tell them that there is a way ahead. No sooner had we announced our presence with barking dogs and yodelling cries than they ran together; but instead of heading off along the steep face they wandered aimlessly about, some climbing up and others down. We could see it was no use relying on them to guide us through the maze—they had come in from our end and they knew no other way out. It was up to us to choose their track for them, but it was soon clear that we were not very good guides.

Whenever we chose a path we found it ended in a sheer rock wall or a platform from which the sheep could not get down or a chute of sliding shingle which they would not cross. The mob began to split and separate—two would go up and several others down. We also separated, following different little mobs and trying to find a track where our clumsy boots could get a foothold. Every now and then our dogs would get above us in steep and broken rocks and dislodge a shower of stones that came bounding down, hurtling past to crash into the trees and alpine scrub beneath us. More and more the sheep broke up into little groups which couldn't or wouldn't join together again, and we began to leave odd ones here and there when they ran into some inextricable place. Trying to poke them out with a stick is often dangerous for they may jump or fall and take you with them down the rocks. Before very long we were almost more concerned to find a way out for ourselves than for the sheep. We could have climbed out on to the top and walked safely along the Mid-Hill ridge but this would have meant abandoning the sheep altogether and we were not yet ready for that confession of failure. We struggled on, the autumn sun beating directly on the rocks around us and making the sweat run down our faces.

All things have an end and after several hours of this continuous struggle we emerged from the last bluffs and stood on the low shoulder beneath the top of Mid-Hill. In front of us, crossing the big shingle basin where the tracks of other sheep were clear and easy to follow, were the pitiful remnants of our original mob, of which no more than twenty had survived the traverse. Reg and I sat down with some relief. It was now late in the afternoon and there had been times when we wondered whether we should get out of that inhospitable maze before darkness made escape impossible. We looked down the long and fairly gentle slope which ends in the Lagoon Saddle—the watershed between the Waimakariri and Rakaia Valleys, where the water which seeps out of the swampy ground gathers in still shallow pools and begins to run downhill, some north some south, to join its chosen river. Reg had his glasses out at once to look for deer, as the whole of Mount Bruce and the Burnt Face was spread out below us, but he could see nothing—which was just as well perhaps! To our left the great serrated chain of the Shaler Range stood out, dark and menacing against the sky, for the sun was dropping down behind it now and from where we sat only Mount Rolleston and the peaks of the Polar Range behind Arthurs Pass were still in sunlight. Hastily we picked up our sticks and began the long descent, knowing that dusk would be on us before we got off the hill.

Reg Ferguson kept cows at the Bealey Hotel, and milked them rather casually whenever he had time. Often when he spent a day mustering for us he would get home late and fail to find his little herd, which had free range up and down the road. There was a paddock across the road opposite the hotel and another tiny one on the higher ground towards the west, but the fences of these were derelict and the cows tended to wander where they would. Unfortunately they often parked themselves in front of the hotel and thereby hangs a distressing tale. Reg's wife, Thelma, grew sweet peas against the front of the pub, and cows are no respecters of gardens, as everybody knows. They ate the sweet peas once; they ate them twice; and then Thelma rang Mary and said, 'You tell Reg that if he comes home tonight I'll shoot him!'

Well, no doubt she had cooled off by the time Reg got home and so no shots were fired that day but when the third and last time came she shot the cow! Not stone-dead on the doorstep of course—she only had Reg's .22 rifle, and the cow, warned by her inarticulate cry of rage, was walking away and presenting only a tail target. The bullet hastened its flight, and tail in the air it tore up into the little paddock on the hill.

This animal was a heifer not yet in milk and so Reg did not miss it at once—only wondered vaguely where it had gone. In fact it never returned from the paddock on the hill; some time within the next day or so it died.

Thelma found it. Perhaps she had a premonition of disaster and went to look, but once found, her one thought was to destroy the evidence of the crime. There seemed to be only one possible method—to bury it, and she had only one person she could trust to help her. Reg's sister was staying at the pub and together the two girls set out armed with picks and shovels in the hope of getting the evidence safely underground by the time Reg came home.

A cow— even a maiden heifer—is a very large animal, and its legs stick out uncompromisingly from the four corners, making it half as large again when you have to dig a hole for it. The Bealey Hotel stood upon a promontory of rock which juts out into the Waimakariri and the task they set themselves was probably impossible. It didn't take them long to decide it was too much for *them* and down they sat beside their monstrous corpse and dissolved in floods of tears.

The epilogue to this sad story is that Reg didn't find the animal until decomposition had destroyed the tell-tale bullet hole, and he never did discover that she hadn't died of tutu poisioning. He was probably the only person in the district to be so ill-informed!

Among the more amusing events of 1944 was my encounter with one of the most notorious characters to have haunted Australia and New Zealand in recent years. We had been haymaking with the help of a gang of young soldiers who had been based in the country districts by a benevolent government to help short-handed farmers with essential harvesting work. There was a camp at Sheffield and from it our gang was drawn. They were delighted by the change from an official army camp to the freedom of a station, where food was plentiful and discipline non-existent, and they tackled the haymaking with energy and goodwill.

Mary had taken the four children that we now had to Sumner for a holiday by the sea, and when the haymaking was finished I was to follow. It went on longer than expected, but when the last bale was made and carted, I got a lift to Sheffield with the returning soldiers in the hope of getting a car or a train which would take me on to Christchurch, since Mary had our car.

There was no train—not even a goods train—for many hours. Petrol rationing had reduced the road-users to a tiny trickle and there was nothing moving in the direction I wanted, so I went across to the hotel to see if a passing motorist had pulled in there for a drink. There was a car outside, I noticed, and when I went into the bar I found quite a few men in there. Most of them were locals—railway men or farmers or Ministry of Works employees, for it was about knocking-off time. Among them, however, were two strangers, the occupants of the car outside. They were well in liquor and chatting gaily with the locals and the proprietor behind the bar. I ordered a beer and waited for an opportunity to ask them if they were going on to town. It soon became apparent that one of them was a doctor, for he was offering examinations to everyone who had a medical problem. Several took advantage of his services and got some rough and ready advice after signing a form which entitled the medical practitioner to seven shillings and sixpence (75c) from the State for each consultation. It all seemed rather unorthodox, and the doctor's obvious state of elation threw considerable doubt upon the value of his prescriptions.

When he had exhausted the potential patients in the bar the pair prepared to make a move, so I approached them and asked if they were on their way to Christchurch, explaining my predicament.

'Oh,' said the doctor, 'we're not going that way—we're going to Oxford. I'm the locum there while the doctor is away.'

A very odd locum too, I thought. However, with the cheerful helpfulness of a half-drunk man he went on to assure me that there was a bus from Oxford to Christchurch every Friday night at six o'clock; he would be happy to drive me there and there would be just time to catch it nicely.

I had my doubts about his driving, but I didn't want to wait in Sheffield till midnight to catch a train, and the bus should surely get me to Christchurch by half past seven so I decided to risk the ride. It was not long before I found out what a risky ride it was. We tore down towards the Waimakariri Gorge bridge, swinging wildly on the sharp corners where the road drops down the terraces above the river, and skidding in loose shingle where it had not been graded. I discovered from the shouted conversation in the front seat that the doctor's companion was some sort of salesman for a drug company, and the frequency with which he addressed our driver as 'Doctor' made me wonder if there was not some deliberate irony about it. After a hair-raising drive, which I think we only survived because the car would not go any faster, I was deposited outside the old East Oxford Hotel from where the doctor said the bus would leave. Alas! the bus was as fictional as the 'doctor', and I had no option but to spend the night in the spartan accommodation provided by the pub and catch a morning bus next day.

It was not till a few weeks later that the scandal hit the newspaper headlines. The 'doctor' had absconded with the real doctor's car and had been arrested on the West Coast, where he was discovered to be an unqualified imposter called Murray Beresford Roberts. It is true that he had some medical training, enough to pose successfully as a locum tenens for Dr Minty, who was completely taken in. He left behind him a catalogue of local stories, which probably had varying elements of truth in them. The best of them perhaps concerned an old lady who had been treated by Dr Minty for some ailment which had remained chronic. When she consulted 'Doctor Roberts' he prescribed some medicine which she was convinced achieved a cure, or at least a big improvement. When poor Dr Minty returned to the shambles of his house and the decay of his practice he found his patient indignant because he could not reproduce the wonder drug which that 'nice young doctor' had prescribed.

Murray Beresford Roberts went on to become legendary in many countries, an imposter incorrigible as any drug addict. He died in poverty not long ago after writing his own version of his many impersonations. What an actor the man would have been if he had chosen to live honestly!

Chapter 6

THE WINTER OF 1945

> Grasmere, Aug. 7th.
> Mr McLeod: Just a note to tell you that there 8 inches of Snow on our selection here. No Tele Phone. Still snowing at 11.30 A.M. you might get home tomorrow and big might at that.
> Yours Faithfully Jack
> Latest Bulletin 12.30 snowing heavy

STATION COOKS have their uses and can be faithful helpers in spite of the stories I have told about them. Jack was best remembered for his 'three-weekers'. He only baked once every three weeks and by the end of each period one of his scones would have been lethal if it had been thrown at a dog!

Many people in Canterbury will remember the year 1945 for the heavy snow which fell in July because, unlike the snow of 1939, which came from the nor'-west and was confined to the back country and the foothills, this one was a true sou'-west snow which blanketed the whole province. Long before that, however, it was a year of difficult conditions. Shearing at Grasmere went on with a scratch gang in very bad weather till late in February, and the continuous wet and the disruption while we were shearing the ewes led to a great deal of scouring among the lambs and an epidemic of fly-strike which prevented us from getting the mobs back on to their own country. The longer we kept them hanging around the paddocks the more of them got struck by the fly, and anybody who has had to deal with that cruel scourge knows what it is like to spend day after day on the stinking job of cleaning up miserable little lambs half-eaten by a seething horde of maggots.

Shearing was so late and rank feed was so plentiful everywhere that we decided not to put any sheep out on Powers Country, so for once we had no autumn muster on that exhilarating country. It would probably have been better if we had, because the rank feed which resulted from the constant rain did the sheep more harm than good and only the cattle throve. At the beginning of the season I had engaged a new tractor driver—a man who was to be with us for many years. His name was Tom Gibson and, though well past the age for service overseas, he had joined the Air Force as ground-staff till he was released. He was an old teamster trained in the traditions of the six-horse team and he never really became at home with tractors; but like most of the old horsemen he knew his implements and could set a plough to make good work which is more than many tractor drivers ever learn.

One day I had a message from Sam Robertshaw, the railway ganger, to tell me that some of our cattle which were grazing on the Ewe Country near our boundary with Craigieburn had been on the line, and one of them had been hit by a train and had a broken leg. The railway gang had put them off the track, said Sam, but I had better go and do something about the injured animal.

Such cattle are always useful for dog tucker so I decided to take Tom with me to help me skin and cut it up after I had shot it. Our two boys were then four and two and always ready for a trip with Daddy in the truck, so they were included in the party as well; Mary was only too glad to have them out of her way for an hour or two. We drove down the Craigieburn Road and soon saw the mob of young cattle which were grazing on the long flat which ends at our boundary gate. I had brought a heading dog to hold the mob together while I singled out the one to shoot, so we parked the truck on the road and all got out. I walked out on to the flat and sent out my dog to gather the thirty-odd yearlings together and bring them down to where we stood. Tom had taken the small boys' hands and led them out on to the flat so that they could watch. The dog went round the scattered cattle and they all ran together down the flat towards me where I stood about a hundred metres from the road. They came fast and it was soon evident which was the injured beast for it lagged a long way behind the herd. Instead of stopping when they saw me in their path they swung round behind me and went down the flat towards the boundary.

I stood there with the rifle cocked and ready, waiting for the injured one to follow, hoping it would stop when it saw me in its path and give me a good chance to shoot it clean. On it came, making surprisingly good speed in spite of one foreleg swinging from the break below the knee. Instead of following the line of the others who had passed behind me, this one chose the path between me and the road; and right in that path stood Tom Gibson and the two little boys. Not only was it heading straight for where they stood but, I realised with a flash of horror, it had lowered its head and was actually setting out to charge the little group. There was no question now of hoping it would stop; I had but one brief chance and I must take it quick.

These are the moments when it is so hard to bear the rules in mind—to control spasmodic movements and to concentrate on careful aim and squeeze instead of pulling at the trigger. The shot rang out and to my unbounded relief I saw the beast stumble and pitch forward, rolling head downwards until it came to rest not five metres from Tom. There was little he could have done. As he said afterwards, 'I had made up my mind to throw the boys down and lie on one of them but I couldn't lie on both!'

We were both shaky after the experience but the boys were too young to have realised their danger, and the whole event to them was just another of those exciting expeditions which made up the best part of their lives.

We got a new shepherd that winter. Ernie Percy, who had been with me since Peter Newton left, was leaving and I had advertised in Otago as well as Canterbury, hoping to find some young musterer from down that way who fancied a change of scene. I got one too, called Lindsay Houliston, who not only served us well for several years but brought a brother and a cousin later on, so it was a lucky strike.

Lindsay's arrival was a bit unfortunate for him as he came on the 12th day of July, and on the 13th and 14th we had the famous 1945 snowfall. Compared to the plains and foothills we escaped fairly lightly, for there was only a foot at Grasmere and, being a south-west snow, it did not deepen much higher up; but it was followed by three weeks of some of the severest frosts we ever had. Lake Pearson froze all over—even Lake Grasmere froze, except just along the southern shore where the springs came in. Magpies and other birds froze in the branches of the trees and fell stark and stiff upon the ground, and it was impossible for the hoggets we had on turnips to get through the icy blanket and get a feed. Luckily we had plenty of hay and a large enclosure round one big barn with hay-racks which we filled up every day. My only concern for the hoggets was for water; there was no water in the enclosure and a diet of dry hay seemed certain to require a little liquid supplement.

We made a snow plough and dragged it from the hay barn to the water race and back again and then set out to drive the mob of hoggets to the water. They had no desire to leave their well-tramped enclosure and embark on a pointless expedition along a narrow and badly graded track with deep snow on either side, and it took us hours to force them in small cuts of a hundred or two along the half kilometre of track. At last the tight-packed mob lined the edge of the narrow stream. There they stood, bored and uninterested in the whole proceeding, while we recalled the age-old saying about leading horses to the water. I can assure anyone who is interested that the aphorism applies equally well to sheep; not one solitary sip would any one of them take and we had no choice but to drive them back to the enclosure whence they came. Whether they got their moisture by licking the dirty snow beneath their feet, or whether they simply did without, I do not know to this day but they survived three weeks on a diet of dry hay and emerged little the worse from their privations.

In the intervals of putting out hay we rushed off in the truck to Cora Lynn to see what the Burnt Face and Bealey Spur were like. The steeper slopes were opening up where they faced the sun and there was nothing like the depth there was in 1939 once you got off the flat, but we had to climb up through the bush track behind the Cora Lynn homestead and rescue over a hundred wethers which had got trapped up near the fence which divided the Burnt Face from the Cora Lynn Tops. The walk up through the bush was sheer delight. There had been heavy frost and no wind ever since the snow fell, and every branch and every twig was snow-laden. The branches drooped beneath their load and the snow itself consisted of enormous crystals which scintillated like diamonds in the sun. There was a silence over the whole world, such as is only possible when it is blanketed with snow. Not a bird twittered and even the whispering streams were silenced by the frost which stilled the movement of all water.

The sheep were grouped in little clusters above the large tract of snow grass which marked the upper limit of the bush before the face was burnt. The pale, feathery leaves stood out above the drifted snow which filled the gaps between the bushes and deterred the sheep from making any attempt to come down. On the dry hard ridges up above they had tramped themselves small areas of standing room from which they looked down hungrily at the clearing spurs below. The danger was not only of starvation if the thaw did not come soon but of murder from above if the keas decided to attack.

Our first task was to collect them in one mob and then to tramp a track through the snow grass on to one of the steep ridges which led to the bush further down the face. This was hard work, as we often stumbled through the soft, drifted snow into holes and hollows in the rough ground among the tussocks; but once we were through and had climbed back up to the sheep it was not difficult to find a leader who would plunge and battle his way to safety while the others followed, urged on by the noisy barking of the dogs.

So we went round the place, snow-raking first on the Bealey Spur and then nearer home on the Long Hill and Bailey behind the house. Here we found the tell-tale traces of keas who had killed two or three hoggets from a snowed-in group in very sight of the homestead. The whole of the east face of the Cass River valley was a sheet of white, and there were little groups of hoggets here and there all over it. There was little we could do for them because there was no clear country at the bottom and it would have been impossible to move them round the face to the front where the steep faces had opened up. At least we could try to protect them from the keas, so Lindsay and I decided to try a night vigil. Loaded with thermos flasks and sandwiches besides the essential weapons, and muffled in thick jerseys and woollen underclothes we set off up the creek. A couple of kilometres or so upstream a razor-edged spur comes down beside a big eroded bluff which scars the mountain side from top to bottom. Half way up is a patch of remnant bush, for this face had been burnt many years before, for what ridiculous purpose it is hard to see. Most of that face of the Bailey range had grassed over quite satisfactorily, but this great scar was no doubt there before the fire and would never heal because its edges continually crumbled. It probably contributed a large proportion of the rocks and shingle which moved down the Cass River during every flood.

We decided to climb up the razor-back and build a fire in the fringe of the bush in the hope of attracting the nocturnal predators. We were opposite the spot on the other side of the valley where keas had killed a ram and several ewes in 1939 and Brig Ferguson and I had made a midnight expedition and shot five around the beleaguered survivors.

Thick clothing is absolutely necessary if you are going to survive a winter's night at 1000 metres or more, but climbing a steep and rugged hillside in the same garments is a different matter. The sweat was pouring off our muffled bodies by the time we reached our goal and Lindsay admitted that the spur was steeper and rougher than any he had encountered in his mustering further south. At last we settled on a suitable spot where there was plenty of dry firewood underneath the trees and where it was open enough for our firelight to be visible all up and down the valley. There we relaxed and watched the golden flames leap up and fling their showers of sparks to the tops of the birch trees, lighting up their dark green foliage and making the floor of our little clearing gleam dazzlingly white. Slowly the heat of our bodies subsided and we became chilled where the sweat had dried on our backs and roasted on the fire-lit side, so that we had to turn, like joints upon a spit, and cook first one side and then the other. It was a great opportunity to get to know a new man, and knowing your men means a great deal when you live and work together as closely as we do. I heard about the places Lindsay had worked on in the south, and the owners of them and their foibles and their good and bad points. It surprises many people to find how much high-country men know about each other even when their stations may be 500 kilometres apart; it is because the men who muster on them like to move about and see new country, and they tell their stories as they go, like medieval troubadours. So we know many men by their repute, and the gossip of Queenstown finds its way up to Hanmer Springs, bringing all high-country men together in one great brotherhood.

So the night wore on. The Southern Cross rode high above the Cass Saddle, and the great constellations wheeled above our heads, but never a kea came down to imitate them. Occasionally we heard the unmistakable cry, but it was from far away across the other side of the river and the bright blaze we kept going didn't succeed in drawing the birds any closer. In the end we gave it up and stumbled wearily down the hill to where the river chuckled at the bottom, laughing perhaps at our discomfiture.

The frozen lakes gave us some entertainment during that winter, and on one occasion a nasty fright. The ice was about 12 centimetres thick and when the sun came out during the day the expansion of the surface set up an intolerable strain, causing great rending cracks to split the sheet from bank to bank and echo round the surrounding hills with an almost musical sound. Some people call this the singing of the ice. The severance is complete, and not until the healing frosts of night fuse them do the floes knit together again.

One day we took the children to Lake Pearson to slide upon this new strange playground while Mary and I skated. The eldest was seven and I was pushing her on a wooden chair which we had brought to sit on while we put our skates on. Faster and faster across the shiny surface I pushed the delighted child—when suddenly we came to one of the open cracks which criss-crossed the 2-kilometre-wide lake. Our weight depressed the ice as we approached and suddenly there was a horrible crash and child and chair dropped from in front of me. My first thought, of course, was that we had gone through into the deep and icy water but I soon realised that all that had happened was that the legs of the chair had struck the sharp edge of the opposite floe and snapped off like carrots, throwing the unfortunate child on to the hard ice among a clutter of shattered wood.

Several years later a much more dangerous accident occurred on Lake Sarah which, being more shaded and shallower, freezes much more easily than the other lakes. It was our youngest son, Robin, who was the victim on this occasion and again a chair figured in the event. There was a pair of black swans upon the lake which had had a late brood of cygnets. These four were too young to fly when winter came and they swam round the lake until the encroaching ice confined them to a small hole a metre or so across; then they managed to scramble out on to the ice and when we came to skate we found them flopping round helplessly, unable to walk on its glassy surface. I caught them one by one and tied their legs together, intending to take them to Lake Grasmere, where as usual there was some open water near the shore.

Leaving them tied I skated off to the far side of the lake and was almost as far away as I could get when I heard a faint crash and a splash. Robin—very unstable on a pair of skates—had pushed the chair to which he clung right over the hole which the young birds had kept open. Instead of 10 centimetres of ice like that all round it was covered by hardly more than 1 centimetre, and Robin just disappeared. Fortunately he kept his hold of the chair and fortunately the chair remained on top of the stronger ice. There was not a moment to lose; if he struggled he might pull the chair in at any moment and anyway nobody would survive long in that freezing water.

I set off across the lake as fast as I could skate, shouting at the top of my voice, 'Don't move! Don't move! Don't move!' Mercifully he kept his head and mercifully again the ice was so strong around the hole that I was able to skate right to its very edge and drag him out.

He was speechless with shock and cold; we bundled him into the truck in silence and Mary set off to drive him home. She always recalls how she breathlessly rushed him into the house, stripped his clothes off while the bath was running and got him thawed out in the hot water and dressed again in warm dry clothes. All this time this normally loquacious child was so shocked that he said not one solitary word. When at last he was dressed and she prepared to drive the truck back to fetch the rest of us he suddenly unlocked his frozen tongue.

'Can I drive the truck back?' he said eagerly.

Then she knew he was all right!

Station cooks have always been difficult to obtain on a back-country station and during the war the problem was more acute than usual, which was saying a lot. In 1943 when Robin was born we acquired one of the best of this rare breed of men. His name was George de la Roche and the first time he came he cooked for all hands in the house instead of in the cookshop. By this time we had installed the big Esse cooker which was one of the best investments we ever made. Unfortunately George hated it because he couldn't see and hear things cooking in it. Shut tightly in heavily insulated ovens the pots could boil merrily away; but unless he could hear the bubbles and see the steam he wouldn't believe it. He was forever opening the oven doors to see if things were cooking or putting more fuel on the fire to try to make it go faster. 'Missus,' he would say, 'I cannot cook on this bloody piece of furniture!'

He *could* cook on it, all the same, because he had the continental flair for cooking, and he taught us many useful tricks about cooking for large numbers. He was stout and cheerful and one of our favourite memories of him is of his figure standing in front of the 'piece of furniture', stirring a pot with one hand and holding the baby under his other arm, partly supported by his ample stomach, which bounced up and down as he ladled and laughed, and probably contributed to Robin's obvious enjoyment.

'Thees ees my little Tittle-Tattle,' he would cry delightedly.

George had a ready tongue and a great sense of humour. One day a girl we had staying with us was helping him dry the dishes and she handed him back a small enamel jug which still had some gravy round the spout.

'You'd better wash that again, George,' she said.

George turned to her quite solemnly and said, 'I have two joogs. Won ees for gravy—the other ees for costard; I always leave a little bit of gravy in won and a little bit of costard in the other so I know vitch is vitch!'

I never knew how much to believe of his life story. He told me he was the son of a Belgian consul in Valparaiso, and spouted a few words of Spanish to prove it. He also claimed to be a cousin of Mazo de la Roche, the well known Canadian novelist, but I think that story is even more unlikely. He said he ran away from home and went to sea and landed in Australia. Here with some money in his pocket he stayed for a while in a boardinghouse, not knowing quite what to do next. In this place one of the rooms was occupied by a youngish woman, and George conveyed the impression that he was a shy and innocent young man deliberately ensnared by a designing female. Perhaps it was so; at any rate she managed on one occasion to lure him into her bedroom and he used to say, 'Dere she vos, sitting oop in bed all poofed oop!' This was accompanied by appropriate gestures with his hands to show the parts that were 'poofed oop'. Things must have made some progress between them because she finally said to him, 'George, what about you and me going into double harness?'

George didn't know what she meant, so the first chance he got he asked his drinking companion in the pub what 'double harness' meant.

'Oh,' said his mate, 'that means getting married.'

George shot out of that bar and up the hill to the boardinghouse and packed his sea clothes before his lady love came home. He was lucky enough to strike a ship which was a man short, and was out of Newcastle that very night. No 'double harness' for him at that stage of his life!

He did marry later on but when we knew him he and his wife had been separated for many years and I don't know how long they had lived together.

The second time George was with us he cooked in the cookshop. At the time somebody had a litter of pups which were housed in a wire-netting enclosure at the back of the stable. They very soon got big enough to scramble out of it and run about the place scavenging and playing. One thing they quickly discovered was that George was a friend, and it didn't take them long to learn when to come for their tucker.

We used to have an old waggon tyre which hung beside the kitchen door instead of the usual disc-blade gong. It had a loud but melodious ring and at the first stroke George made upon it each pup, wherever he might be, would set up a wild shrieking and yapping and set off as fast as his little legs could carry him in the direction of the cookshop. They all converged at about the same time and submerged George to the knees in a seething mass of black and white and tan bodies. George would stand there gazing down upon them beneficently over his aproned paunch. 'Dem fellas'll eat me ven dey get a bit bigger,' he would say.

He used to suffer from what he called malaria, pronounced with a strong accent as if there was a guttural 'ch' before the 'r'. Brandy was his sovereign cure for it and it was his favourite drink at any time, though he was not above drinking any form of alcohol if he could get it. He had cooked several times at the Ball Hut during the skiing season and he told us how he used to pretend to be in a violent temper so that Vigley, as he called his boss, would bring him some beer to pacify him.

George's 'malaria' was not what he believed, or if it was the cure was worse than the disease, because his attacks were mainly heart trouble as it proved in the end.

He was one of those who always came back from town on the day he said he would; he might not be in very good condition but at least we could be sure he would be there. After the snow in 1945 he had gone down to the races, and on the day appointed for his return there was no sign of George. We didn't do anything about it because we were sure we should hear from him. However, a couple of days went by and there was no sign of him and no word. After nearly a week the police rang up and said that a man called de la Roche had been found dead in a boardinghouse in Christchurch, and that he appeared to be employed by us. I said that was so and that I was very sorry to hear of his death.

'Well,' said the detective, 'could you come to Christchurch and identify him?'

'Hasn't he any relatives?' I said. 'I believe he has a wife somewhere. Why pick on me?'

'The fact of the matter is,' said the detective, 'that his wife says she hasn't seen him for so long that she doesn't know what he looks like!'

In the end I went down and presented myself at the hospital, where I was introduced to the coroner. He was a tall, gloomy man with a profound sense of the solemnity of the occasion. He asked me if the deceased was a close relative. We marched in silence through dark corridors, trying not to disturb the occupants of the mortuary with the tramp of our booted heels. I have the impression of a large, low-ceilinged hall occupied by a number of discreetly sheeted biers, to one of which we were led by an attendant. The sheet had been turned back to reveal a still, unsmiling face, recognisable as George, but without all that subtle force, sly humour and sheer zest which made his personality when alive. Lower down, the sheet was raised into a large hump by that fine, Falstaffian stomach, never more to heave and bounce with Rabelaisian mirth.

We turned away silently and indifferent hands covered his face.

Next day, again alone, I buried him, riding, to save wartime petrol, in the hearse with one lone wreath and the undertaker for company. We carried him, the undertaker and I and the graveyard staff, and he was no light weight. Over his head a Catholic priest said the Latin words that sound so strange outside a schoolroom.

When I got home Mary asked me who I got to bury him and when I said I thought he was a Catholic she howled with laughter and said that George had often told her what he thought of the Holy Church of Rome. Well, George would have seen the joke and perhaps if his soul was hovering near, it wheezed and spluttered as he did in life to see the church he had forsaken accept his body when he had no further use for it.

George was old and he died as he would probably have wished, before decay began to make life a burden to him.

George's death was not the last of the tragedies of 1945 and the final one was a grim drama which none of us will ever forget. It formed another tragic chapter in the long history of the Bealey Hotel. We started the tailing muster at the beginning of December and had moved up to Cora Lynn to camp at the old homestead to muster the ewes on the blocks up and down the river. Reg Ferguson was helping us as usual and one day after an early start we finished tailing about midday and retired to the cottage to eat and doze away the afternoon. Reg, of course, went home to the hotel. Not for him the idle afternoon that satisfied the rest of us; early summer is the time that deer are most frequently seen in the open since there is by now plentiful feed about the lower country and there is little temptation for them to move out into the high basins where patches of snow still cling to shady spurs.

Half way through the drowsy afternoon we were roused by the sound of a car and the next moment a man charged into the house with an incoherent story that Reg had been shot while deer-hunting on the Bealey Spur. The face of the hill behind the hotel is very steep, with outcrops of solid, ice-smoothed rocks standing out grey and white among the dull green scrub and tussock. We climbed it at breakneck speed but there was little point in haste, as the first brief glance made clear. No man could help him now.

The story of that fatal afternoon revealed itself slowly, as official enquiries and private stories unfolded bit by bit.

There were always deer about the upper part of the Bealey Spur, where scattered clumps of bush surrounded a small string of shallow lagoons and large patches of tall, waving snow grass made good cover among which their rusty flanks were difficult to see. Not content with a long day in the open tailing lambs, when Reg got home he decided to walk up the hill behind the hotel and see whether from the eminence of one of its shoulders he could see deer, either on the spur itself or up and down the wide Waimakariri river bed spread out beneath his feet. His wife often went with him on these expeditions; she had her own rifle, and on this occasion they set out together.

This, briefly, was the story that she told. The hill, as I have said, was very steep and Reg's long legs and strong muscles soon carried him ahead. He reached the top of the first shoulder while she was still plodding slowly up with her rifle slung across her back to leave her hands free. He sat down on a low, rocky knob and took out his field glasses, searching the country carefully, segment by segment, while she completed the climb. Breathless and with shaky legs she walked past and seated herself on another rock a metre or two behind her husband. Then, intending to disencumber herself of the rifle on her back, she leaned forward to lift it over her head. Whether it was cocked—whether the safety catch was on or off—whether the trigger caught in a fold of her coat—these questions were never satisfactorily answered. The terrible fact alone was clear; as the muzzle of the rifle dipped towards the man in front it went off and the bullet struck him at point blank range in the back of the head.

Of all the ghastly situations for a wife to find herself in this one would be difficult to equal and it is no wonder if her first story was incoherent when she staggered exhausted into the hotel some time later.

The tragedy hung over our work during the next few days.

The gap in our little rank at the tailing yard was painfully noticeable and when the funeral took place at the little Springfield township a few days later it was very obvious what a well-loved figure he had been. Not only the people from the district, where the Fergusons had lived for years, but many who had become his friends from staying at the hotel and enjoying his company on shooting expeditions gathered for their last farewell. The little church was packed and a crowd surrounded it outside while the service went on; but it was not until the final hymn that the drama reached its climax. The hymn *Abide With Me* is charged with emotion at any time, but under these circumstances it was unendurable. As the voices took up the words of the last verse the poor widow's self-control broke down completely, and far above the sound of the music shriek upon hysterical shriek rang out from the front of the little church; and while the cortège moved down the aisle and out between the silent mourners outside, the despairing cries still rang in our ears. I for one, of that sad gathering, can never hear this hymn without recalling in vivid detail the scene in the church that day, nor the bitter emotions that swayed the watching mourners.

Chapter 7

COOKS AND THEIR WAYS

God sends meat; the devil sends cooks.
(ENGLISH PROVERB)

ONE OF OUR MOST amusing interludes—and fortunately it was only an interlude—was our encounter with an old character called Joe.

It was wartime and as I have recounted, cooks were terribly hard to get. The children were young and the work of looking after four of them and cooking for men was just too much. Money was short because the wartime price of fine wools was fixed at the low figure of the late thirties, so we wanted a man who would cook in the house to save the cost of keeping the two kitchens going.

Now many cooks don't like cooking in the house; they have to conform to standards of tidiness and hygiene which are irksome, and they often have to provide a variety of food beyond their usual range. They also dislike being ordered about by a woman and prefer the independence of the cookshop.

We advertised for a cook and were lucky enough to get one reply. This was from a man who wrote from an address in Christchurch and claimed an impressive list of past experience. We decided to grab him while we had the chance, and as we had to go to town we arranged to pick him up. He looked rather old, but we were most impressed with a small parcel which he said contained a new cap and aprons which he had specially bought for the job. 'Ah!' we thought, 'Here's one who really takes a pride in his job and means to stay'. The date, which has some bearing on this tale, was 15 July.

Next day Joe appeared, not only in a clean white apron, but in a tall white hat which would have done credit to the head chef at the Savoy Hotel. We looked at each other with wonder and anticipation; could we really have found the long-sought treasure?

Heat-storage ranges are quite common nowadays but they require to some extent an adapted technique in cooking. A slow fire smoulders day and night inside a massive and well insulated box. A small draught passing through tiny passages carries the heat round the several ovens and finally up a 10-centimetre flue, by which time most of the heat has been absorbed by the ovens and little is lost to the outer air. On the top, instead of a wide expanse of iron of varying degrees of heat there is one hot plate or sometimes two, covered, when not in use, by insulated lids. These hot plates lose heat very quickly if left uncovered, and instead of boiling all vegetables and other pots on them the system is to bring everything to the boil and then put it in one of the ovens to go on cooking. Even George, who was a man of active mind, had found this change too difficult for one of lifelong habit and it soon became obvious that to deal with such a strange idea was quite beyond Joe's capacity.

We explained it all a number of times and Joe would say he understood and that everything would be quite all right. However, what he used to do was to have a group of pots on top of the stove, surrounding the hot plate, the lid of which was firmly down. As soon as he was sure nobody was about he would lift the lid and have the pots all boiling merrily on the plate. Directly he heard anyone coming, down would go the lid and the pots would stand innocently on the stove waiting, as he would say, to go into the oven.

The result of all this was of course that the hot plate, and finally all the ovens, would get cold as a result of the dissipation of heat; dinner would be late, the meat half-cooked and Joe distraught and angry.

One of his reasons for refusing to put pots in the ovens was his inability to find the ovens. In our stove there are three main ovens and the fire box, all with identical doors, and on the left a huge 'auxiliary' oven with a door the full height of the stove, in which a whole dinner with its plates and dishes can be kept hot. The three main ovens are of varying heat to suit the needs of different dishes.

Joe would stand before this monster with a puzzled expression on his face. Tentatively he would open first one oven and then another, searching for the elusive pot of potatoes which he knew he had put somewhere in the stove's recesses. Sometimes after a process of elimination he would triumphantly fling open the only remaining door to find to his disgust nothing but the fire box inside.

'Keep calm, Joe,' he would cry, clapping his hand to his aching brow, 'it must be somewhere!'

We soon found that the tall hat and the lots of vast experience were only the cloak with which he tried to hide his complete ignorance of cooking. I always say that Joe had cooking down to the utmost simplicity which can be achieved—he had in fact one recipe! I have seen in cookery books a standard recipe which can be used for most small cakes, biscuits etc. and varied by flavouring and the insertion of different kinds of fruit. Joe extended this to everything except meat and vegetables. If he cut it into small pieces and baked it, it was scones; if he put some currants or raisins in, it was little cakes; if he put it in a large tin it was cake; if he rolled it out and put it on top of a pie dish, it was pastry and—so help me, Heaven!—if he boiled it in a cloth it was duff! I hardly need add that whatever he did it was either unpalatable or downright uneatable! When it was obviously uneatable because everyone left it on their plates Joe had one easy way out.

'Give it to the children,' he would say, 'it'll be all right for them.'

You can imagine what the children had to say about that!

Fortunately Joe had one virtue—he went to bed early, and once he was safely away in his little room outside, Mary and I would rush into the kitchen and bake shortbread, afghans and chocolate cake and feast to our hearts' content by the sitting room fire.

After three days the puddings for dinner at night had consisted of duff, apple pie, and rice pudding. On the fourth day Joe said cheerily, 'What are we going to have tonight, missus? We've had all the puddings there are!'

At the end of a week we could bear it no longer—we gave Joe a week's wages in lieu of notice and he departed. He didn't seem to mind at all and he left us as a memento the tall hat and all the aprons, a costume he would never want again.

Next morning I heard on the wireless the familiar announcement: 'The Social Security Department states that the opening day of payment of widows', invalids', old age and war pensions will be Tuesday July 22nd.' Then, of course, the penny dropped. Joe hadn't been able to last out till pension day, and had taken a job for a week. If we hadn't sacked him he would have left anyway!

Some time after this we had a cook called Jim. Now Jim was a good cook and even made good pastry, which is much rarer than hens' teeth in station cookshops; not only was his pastry good but he used to glaze it with beaten egg to make it look nice. Unfortunately, like many another good cook, he wasn't over-fond of washing-up—or washing down for that matter either. Cleanliness in fact was his weak point and he much preferred sitting in a chair reading books to scrubbing floors and washing dishes. He was with us for quite a while and even survived several trips to town without mishap.

One thing is anathema to me and that is the man who doesn't come back when he is due; there is nothing more infuriating than to have your plans upset because a cook or any other man just stays away an extra day or two. In the end that is what Jim did, and after giving him a few days' grace we found another cook. Well, you don't like a cook to go into a dirty cookshop, so I went up to see what sort of a state the place was in. Outwardly it was not so bad; the floor was reasonably clean and the closed cupboards revealed no hint of what was inside. I walked into the room and suddenly my nose became aware of a subtle and elusive smell. It was here and it was gone; it was there one moment and somewhere else the next. It was not the rancid smell of stale fat which unclean cooks accumulate after a few weeks; it was more pungent, and it gave you each time that retching feeling that you get from something dead. I thought of meat left in the cupboard, of trapped mice, of dead cats under the floor, but somehow it didn't seem quite like any of these. In the end, as I couldn't pin it down to any place, I decided to go through all the cupboards and clean everything from top to bottom.

Each cupboard yielded opened packets of food of every kind, half-empty tins of jam, old biscuits, stale cake and the endless paraphernalia of a kitchen which is seldom cleaned. In fairness to Jim it's possible that not all of this was his debris, some may have been there before he came.

Methodically I went through the lot—throwing out mouldy jam and mousey rice, washing pots, plates, mugs, tins and everything I could get my hands on. In the last cupboard I found IT and discovered why IT's aroma was so elusive and so penetrating and why I had failed to find it when I looked for something much larger. IT was a tin mug, in which was stuck a worn and almost hairless shaving brush embedded in rotten egg! This of course was the glazing for his pastry and there it had sat for who knows how many weeks, hidden behind a dirty bowl and slowly poisoning the atmosphere.

I had to go away for a few days and while I was away the second act of the drama took place.

About 1 a.m. one night a car drove up and stopped outside the house, and after some time, during which people walked up and down and voices both male and female were heard, a man walked round on the gravel in front of the house. My wife got up and opened the window and asked him what he wanted.

'Oh!' he said, 'I've brought your cook back.'

'What cook?' she asked, for the new one had not yet come.

'Jim Hancock.'

'He's not our cook any more—he's left.'

'Well, he said you owed him some money and he'd pay me when we got here.'

'Nothing doing! He was paid up to date before he went away.'

'But he owes me fifteen pounds!' The voice was getting anxious.

'Can't help that—where did you pick him up?'

'Oh, the Royal Dance Hall. He said you'd pay me.'

'Well you ought to have more sense than to pick up a drunken cook at a dance hall.' The taxi driver turned away disconsolately and the sound of voices rose again in angry altercation.

Mary went back to bed but suddenly remembered the cook's ration book and went to the kitchen, thinking she might as well give it to him. While she was there the door opened and in came our hardened old cowman who had been roused by the party. She gave him the book and then in a spasm of kindly feeling she said, 'Poor things! Tell them to come in and I'll give them a cup of tea, they've been driving almost all night and there are some women there, too.'

'Tea be damned,' said Alec, 'you go back to bed, missus, I'll deal with them.'

'But the poor things, they've been lost, and gone all the way up to the Bealey. I'll give them a cup of tea.'

'You'll do no such thing, Missus. You go back to bed. They've been lost all right!' And he went out, grumbling and swearing under his breath. Finally the car drove off and all was peace again.

In the morning Mary got the men their breakfast. Old Alec waited till they were all sitting round the table, and then, with the dirtiest leer on his weather-beaten old face, spoke up loud and clear.

'Well,' he said, 'I've been on a few bashes in my time but I never brought a couple of harlots with me when I came home. And you,' he said, looking scornfully at Mary, 'wanted to give them cups of tea!'

Poor girl, it had never dawned on her what the occupation was likely to be of ladies who went on taxi rides with drunken cooks in the middle of the night!

There was an epilogue to this story and I leave it to others to decide whether I took legitimate revenge. Not long after this there came a note from Jim Hancock requesting that his swag be forwarded to a Public Works camp further south.

I rolled his blankets and stuffed them and all his gear in a sack; but before I tied the mouth I put in the tin pannikin, which still contained the shaving brush and the rotten egg. I often wished I could have seen his face when he opened it!

Things are often funnier in retrospect than during the heat of the moment. Among our reminiscences of cooks, I have laughed more over the story of Uncle Fred than any other.

We had been going through a very bad period. Cook succeeded cook in quick succession, each one greasier and lazier than his predecessor. As shearing was approaching it was urgent that we find someone who would see it through, even if he left directly after.

At the very last moment a man rang up in answer to our advertisement. He had a broad accent which I couldn't quite identify; either Scots or Irish. He said he had just landed from Australia and wanted a job straight away. When I asked him about his experience he reeled off a list of hotels in Scotland and finally told me he had cooked in the restaurant car of the Royal Scot which runs between London and Edinburgh. What were the inconveniences of a shearers' cookshop to such a man? After producing succulent meals for wealthy businessmen, and sportsmen on their way to the grouse moors, in a rattling compartment the size of a small bathroom, the needs of the humble shearer ought to be child's play to him. So I booked him at once. When he arrived he proved to be no Scotsman at all but a man from Northern Ireland, who had spent many years in Australia. When he had done his service at Gleneagles and on the Royal Scot I never discovered, but I came in the end to suspect that it was in his extreme youth, and strictly in the capacity of potato peeler or some such. At any rate he could leave less of a potato when he had peeled it than any man I ever saw.

His conversation consisted of, 'Oo aye, she'll be right,' a

mixture of British-colonial which meant precisely nothing at all. Beyond that the only noises he made were weird monotonous songs in a language which everyone took to be Gaelic, but was probably only the same phrase set to music!

I was a little alarmed soon after his arrival by his request for more rice. He had already been given a sizeable bag of the stuff, and it was obvious that if he wanted more he must be going to feed the men exclusively on this rather unpopular diet. When I discovered that he had used nearly all the first lot of rice *in one pudding* I became really anxious, and pictured a gang of shearers with ruptured stomachs, because even I know that it only needs three ounces of rice to a quart of milk, and he must have used at least three pounds.

It is not difficult to gauge the standard of a shearers' cook by watching the behaviour of the men. Shearers habitually eat their meals quickly and spend most of their leisure in the shed, but the speed with which the men reappeared after lunch soon showed me that they couldn't be eating much.

When the smokos are brought down at 9.30 a.m. and 3 p.m. the men gather round a bale of wool which serves for a table, help themselves to a mug of tea and then select from a box of food such items as they fancy from its contents. Their actions and expressions as they survey the delicacies provided for them are a wordless commentary upon the cook. If he is satisfactory and the food attractive they take two or three pieces of scone, cake or sandwich, and retire without comment, rest their backs against the wall and exchange good-humoured banter among themselves. But if they start to pick and choose, take a bite and throw the rest out the door to the ever-hungry dogs, and stalk across the floor with that stiff, expressive back which signifies disapproval as clearly as any words, then it's time the boss had a look at the tucker himself. On this occasion the men were unusually patient, and even after several days during which they ate less and less and the dogs more and more there was no outright revolt. They knew I had done my very best to get a cook and so they contented themselves with including me in the general misery as a fellow sufferer rather than a culprit. They even got some amusement out of the situation, particularly if one of the dogs turned up its nose at a particularly uninviting morsel.

I began early in the week to try to get someone to replace this cook but I dared not say much to him for fear he might leave on the spot. At last a woman rang up in answer to my advertisement and offered to take the last week of the shearing and remain as permanent station cook. She sounded as if she knew her job and her own mind and without further queries I engaged her.

When I went to tell the Irishman that I didn't need his services any longer he was quite philosophical; remarked that he was sick of shearers anyway, and departed without asking for a week's wages in lieu of notice.

When I made up his wages form I saw that he described himself as a 'meat worker'. There are many jobs in meat works, some of them as unsavoury as the food he managed to produce.

In the long years since 1930 I have cleaned up the cookshop many times but I doubt if any of his predecessors succeeded in making it so filthy in so short a time. We were hard at work trying to remove the grime and the rancid smell when Gladys arrived. She came in a small and battered car which she said belonged to Uncle Fred who had brought her up and who would stay a couple of days to help her get going.

Gladys was short and neat and clean, with a hard green eye and a clear knowledge of what she wanted and what she intended to have. As a shearers' cook I liked her looks but as a permanent one I was not quite so sure—she was too fond of demanding things. She set to work like a whirlwind and soon restored the cookshop to order and cleanliness.

Uncle Fred fetched and carried and seemed to be a thoroughly nice old chap. I shouldn't like to guess at her age but she didn't look much over forty while he must have been sixty. She was as efficient as she looked and once she discovered the point beyond which demands would not be met she adjusted herself accordingly and things went very well.

As I surmised she was fully capable of keeping discipline in the cookshop, and savoury meals there were followed by tasty sandwiches and small cakes and excellent scones in the woolshed.

Uncle Fred found it convenient to stay beyond the two days first suggested but he was helpful and I saw no reason to object to his presence if she liked to keep him as an unpaid offsider. The weather was good that week and by Thursday night all the sheep were shorn and most of the gang drifted up to the pub for a night of celebration. With them went Gladys and Uncle Fred. Next morning there was the usual cleaning up of the shed, carting of wool and departure of the shearers and shed hands; but nobody thought fit to tell me that they got no breakfast, lunch or tea.

The next day we had to go away and we did not get back until Monday night. Then the storm broke. Nobody had got anything to eat over the weekend except what they had cooked themselves. On Sunday evening our head shepherd had seen a light on in the house and knowing we were still away, he had gone in to see what was happening.

An exploration of the house revealed no sign of any intruder until the sound of stertorous breathing behind the door of the telephone room finally gave the show away. There was Gladys, clutching a bottle of white wine to her bosom and leaning for support against the wall. When she was discovered she described in the minutest detail the ancestry and habits of the loyal shepherd who had unmasked her. What her speech lacked in precision it gained in force and imagination and when he forcibly removed the wine and ejected her from the house her rage was beyond description. I daresay he gave as good as he got and small blame to a man who had spent a hungry weekend as a result of her indulgence.

When we arrived home on Monday night the cat was out of the bag, and even if nobody had told us what had happened we would soon have discovered the absence of two bottles of whisky, one of champagne—left over from our daughter's wedding— and two of good red wine! She had started off with great cunning, walking down to the house with a pudding basin covered with a tea towel as if in the innocent occupation of getting some stores, and returning with the basin still covered with the tea towel but full to the brim with whisky!

What a party she and Uncle Fred must have had! At some stage when the shearers were still there Uncle Fred had fallen on his face. Unfortunately his face had coincided with the bumper bar of a car and he arose with his glasses still on indeed but without the lenses which enabled him to see. Drunk before, he now staggered wildly about bumping into things and people and unable to understand what had happened to his normal vision. Next day of course a large bruise and two half-closed eyes made it impossible for him to wear the glasses at all and for the rest of the time he was with us he was a battered and pitiful sight.

On Tuesday morning I stalked into the cookshop, full of righteous indignation, to be greeted by Gladys, her green eyes bloodshot and her hair dishevelled, staggering out of her bedroom in a dressing gown. Nobody of course had had any breakfast and the stove was dead and cold. Even then she put on a brave show and almost succeeded in convincing me that it was not she who had taken all the drink. One bottle of whisky she admitted but the wide-eyed sincerity with which she denied the remainder might have fooled me if she had not told the man who caught her that she had had two. Luckily for us we hardly ever drink whisky or we might have lost much more.

No formal notice was given or expected and when I came back with her cheque she had packed most of her things and was almost ready to depart. She had one more brazen attempt at a try-on when she said to me, 'Who pays Uncle Fred; is it you or the shearing contractor?'

I said very firmly, 'Nobody pays Uncle Fred. You brought him here to help you, you pay him!'

The joy of this episode is in the scene of their final departure.

When the car was loaded Gladys drove it down from the cookshop to where I waited to see them off the premises. Laboriously Uncle Fred got himself out of the front seat, made his way, with some difficulty, round the car to where I stood, and propping himself against the roof with his left hand he held out the other to me.

'Goodbye, Mr McLeod,' he said, 'it has been extremely kind of you to have me here all this time. I have enjoyed every minute of it!'

No honoured guest departing from my house ever addressed me with more sincerity, or with more correct and appropriate phrases. I had no difficulty in believing that he meant every word he said—the dear old boy!

Chapter 8

JOURNEY INTO THE PAST

> Skirting round the bush with owls still calling,
> We tried to catch the cattle as they lay,
> For once they broke their camp in early morning,
> The wild ones took to cover for the day.
>   E. C. STUDHOLME, *Cattle Mustering*

CATTLE FARMING in the high country has become so popular of recent years that one tends to forget that it has not always been either very profitable or very successfully carried out. The old merino shepherds were often extremely poor hands with cattle; they neither liked nor understood them. The result was that many stations, like Mount White and Mesopotamia when I mustered there, had nucleus herds of wild cattle in some inaccessible valley; and in some cases they became so cunning that they were even difficult to eradicate by shooting. Just how easily this can happen was illustrated to us on what anybody would regard as comparatively easy country.

We had been in the habit of solving the problem of keeping cows and calves apart at weaning by putting the calves we wished to keep in a railway truck, sending them down the line to Craigieburn and then releasing them on the block which we called the Ewe Country, whose boundary is less than a kilometre from the Craigieburn siding. Disoriented by their blindfold journey, they seemed unable to find their way home, and settled down quite peacefully there while their mothers, suddenly bereft of their young, roamed about the country where they saw them last. Normally these calves would be mustered the following spring and removed to a different block; one year, however, I left them there through the summer and it was only when the block was required for another crop of calves that I sent for them. The mustering gang, consisting of three or four men with about twenty-five dogs, were to ride up to Grasmere from Craigieburn and I said to Lindsay, who was in charge: 'As you come up the Ewe Country pick up those yearling cattle and put them across the railway line on to the Long Hill,' knowing that they habitually grazed on the flat beside the line.

When the men got home Lindsay broke the news to me rather shame-facedly that the cattle had broken away from them and scattered all over the hill, pursued unsuccessfully by men, horses and uncontrollable dogs. It was my own fault as I well knew. Cattle mustering is usually done with a circumspection unnecessary with sheep. One or two dogs under good control are all that are needed, and to approach young cattle with a cavalcade accompanied by a veritable pack of hunting dogs is just asking for trouble.

The Ewe Country has a narrow flat on the railway side and three large hills divided by deep saddles which lead through to a steep face which falls right to the Waimakariri River. Once the cattle bolt through one of these saddles it is impossible to chase them on horseback, and on the downhill slope no dog can hope to head them.

It took two years to get the last of these cattle off that block and the hero in the end was Brig Ferguson's old dog, Peel, one of a famous heading breed, who simply trotted ahead of the cattle, right round the block and back to where they started. They could not shake him off and he never really tried to stop them but in the end they were so sick and tired of running that they stopped of their own accord. The incident taught me a few useful lessons; never leave cattle on a block too long and never pass them without putting a dog round and drawing them up. If once they learn to run away when they see a man they will always do it, and few dogs can stop them when they are headed towards a refuge that they know.

This was not the end of our cattle troubles at that time. We were in the process of changing from Herefords to Angus because I thought the latter better climbers which would utilise our steep hill country more effectively. I had bought a couple of trucks of two-year-old black heifers in the autumn to put to the bull in the following spring. They were to winter on the flats at Cora Lynn, but one winter's day when I drove up the river I looked over the high bank where the road climbs round above the river and to my horror saw two unmistakably upside-down black bodies at the foot. Although they came from a place where the tutu plant is common enough, they had fallen a victim to its virulent poison, where it grows among the fern and rocks above the river. This is a strange thing—we never lose cattle bred on our own country but whenever we buy any the same thing always seems to happen—some of them eat tutu and die.

In October 1946 another unexpected death involved me in a strange, nostalgic expedition which carried me right back to the days twenty years before when as a raw young Englishman I began my New Zealand high-country life. We heard one day that Jim Thompson, who had managed Mount White Station ever since I was there in 1926, had died suddenly in the stable yard, and very soon after I had a strange request. There were no men in the vicinity of the Mount White homestead except a cowman gardener and Jim's son Alec—and he had to go to Greymouth for the inquest because by some strange freak of administration half of the upper Waimakariri Gorge is in the Westland Police District. The only two men employed on Mount White Station at that time were at the old Anna Hut, far up the Esk River, 30 kilometres from the homestead. The cowman did not know the way and there was no method of communicating the news to these men except by word of mouth. I was the only person in the district who knew the track, so they asked me if I would ride out and tell the men what had happened. Nowadays there is a Landrover track all the way, but then an ill-marked packtrack was all there was to guide you once you got past Nigger Hill, exactly as it had been twenty years before. I did not know what horses I would find at Mount White and I did not want to lose a whole day by riding from Grasmere so I went to Craigieburn, borrowed a horse from there and set off to ride a more direct route over the hill to the confluence of the Waimakariri and the Esk and up the Esk to Lochinvar. My borrowed horse turned out to be a fair traveller and I found a ford across the Waimakariri without having to swim, which was a great relief because the river can be bitterly cold in October and is always swift. Once up out of the Esk on to the old road to Nigger Hill I was on familiar ground, and the memories came flooding back as the twisting track revealed each creek and spur and patch of remnant bush which had etched their details on my mind the first time I had traversed it in 1926. Then I had been sent with unfamiliar packhorses to carry chaff to Nigger Hill, where a married man, Jack Deans, lived with his wife and daughter in a little house half hidden in the bush.

As I rode I tried to measure the changes which had taken place since that time. Was the erosion worse—the bush more sparse—the shingle lower on the ranges on either side? How did the 'lazy man's beat' on the other side of the Esk compare with the time I saw it last? This was a piece of natural 'badland' which formed a precipitous terrace constantly eroded by the river at its foot and christened ironically by the unfortunate musterers who had to walk through it on foot. All my impressions on the way to Nigger Hill were that the growth was more rank, the manuka scrub more extensive; and the remnant patches of bush were spreading out in a healthy fringe of young mountain beech. This remnant bush in the Esk Valley was the result of one or more enormous fires whose date I do not know. Traditionally the story was that a fire from Lees Valley on the eastern side of the Puketeraki Range came over the top and burnt the large forest on both sides of the Esk. This seems to me to be unlikely, to put it mildly! The top of the range would have been very sparsely vegetated, if not actually bare shingle, and the prevailing winds were in the opposite direction. It seems much more probable to me that the early settlers fired the forest on both sides of the Esk to produce grazing country. At any rate the only extensive area of bush they left was a wide strip in the part known as 'Middle Pakate', which acted as a boundary for sheep grazing below it in winter and above it in summer.

The Deans' old cottage, built of pit-sawn timber from the surrounding bush, has long been stripped of its iron and allowed to sink back into its natural surroundings, and only the tin musterers' hut whose smokey chimney I had good cause to remember remained of the Nigger Hill I knew. From here the packtrack was as ill defined as I remembered it but I needed no signposts on a route which I had walked and ridden many times. When I was last there the tussock was very sparse, for it is a poor hungry glacial plain until you drop again into the valley of the Esk. What I did notice now was a great increase in scrub of every kind; not only manuka as in the country through which I had come, but cassinia, coprosma, and every kind of alpine hebe clothed the once open slopes of Big and Little Flora and the shady faces of 'Main Range', the great spur which divides the upper from the lower Lochinvar Country.

I reached 'Surveyors Knob' which is supposed to mark the grave of an early surveyor who died here—drowned in the Esk perhaps?—and here I was reminded of a very recent mishap, when a training plane of the New Zealand Air Force had lost its way in heavy cloud and been forced to land on these inhospitable flats because it had hardly any petrol left. Jim Thompson had told me how they sent packhorses loaded with aviation petrol out all the way from Mount White, and dug tracks across the rough, tussocky flat to the edge of a high terrace to enable the plane to take off. He also told me of his indignation when the pilot refused to put the petrol in his tanks because he said that now he knew where he was he could reach an airfield on the petrol he had—provided he could take off. If he put the weight of the extra petrol in he might not even get off the ground! One can imagine the tense faces of the watchers, not to mention the pilot, as the plane leaped and bounced across the impossibly uneven ground, and the ghastly moment of suspense when it reached the edge of the terrace. Mercifully its speed was great enough and it dropped away airborne over the deep valley below the terrace and landed safely at Wigram airbase half an hour later.

From this point the track drops down the terrace into the Esk Valley and I was astonished to see on a patch of hillside across the stream a concentration of twenty-five wild pigs. Pigs have always been numerous here and not far ahead was a scrubby spur called Pig Spur which comes off the Pakate Range opposite the Anna Hut. The patch of hillside where this veritable herd of pigs had collected had been burnt not long before, whether accidentally or to clear the scrub for mustering I did not know. At any rate the fresh green growth evidently appealed, not only to the grazing sheep, but to the pigs as well, and they were spread out across the patch of one or two hectares, rooting away busily. There were pigs of every size and colour, black, white, blue and piebald, and the sight of a passing rider disturbed them very little. When I had mustered there twenty years before, the owner, Ronald Turnbull, had always set his face against the destruction of pigs because he said they kept the rabbits down. The musterers thought they kept the lambs down as well, but anyway they were good fun to hunt and extra good as a change from the eternal mutton chops.

The Anna Hut was little changed. It crouched beside a spring stream in a little gully which led into the Ant Creek. Its floor was dirt, its walls were iron and its old timbers had been cut out of the bush I cannot tell you how many years before. The story goes that in 1893 when Edward Chapman was accidentally shot while hunting wild cattle his body was hung up with packstraps to these rafters. One thing was certain—the bunks had not improved since I last slept in them, and I was not sorry that I only had one night to spend under this inhospitable roof.

The men, of course, were surprised to see me and still more surprised to hear my news. Next day we all rode home together.

This expedition gave me a great deal to think about and made me wish I had taken more photographs in my mustering days to show what the vegetation was like then, for it was useless for me to say now: 'When I was mustering there that face or spur or gully was quite clear and now it's covered with scrub.' I believed in my memory, but who else would? Too many mistakes have already been made by people who said the exact reverse, for anyone's memory to be accepted as accurate. The rate of change is always too slow to measure with the eye unless some sudden cataclysm strikes and great yawning guts open on a hillside overnight. I was beginning to realise that in the last forty years or so the lower slopes of all this mountain country had begun to recover from the first assaults of the white settlers as the practice of burning was reduced and the sheep numbers died down from their peak at the beginning of the century. What was happening to the mountain tops was a very different question. There the vegetation had always been so sparse that increase or reduction was very hard to measure. The sheep farmer had one answer and one answer alone. If his sheep came off the tops in the autumn in good condition the grazing must be all right; but the scientific workers of the Forest Service and the catchment boards began to tell another story, and the long struggle began—to establish the system of land classification according to its vulnerability to erosion and to relate land usage to these classifications. Thirty years later the struggle still goes on, and during those thirty years much of my time has been spent trying to ensure that high-country sheep farmers should accept the truth and prosper in spite of it.

Many years later a strange coincidence threw interesting light on the early history of the Lochinvar Country. The Bank of New Zealand in Christchurch's Cathedral Square was being rebuilt, and all the archives from the dusty cellars were being unearthed and transferred to other premises. The manager, Mr Travers, who was a grandson of that Travers, the eminent botanist, who is remembered in the names of many of our native alpine plants, asked me if I could throw any light on a document which he had found. It proved to be a deed, or bill of sale, written in archaic longhand script with all the legal terminology of 1863, securing a loan of £13,000 from the Bank of New Zealand to William Sefton Moorhouse and Richard May Morten, with security over several leases in the upper

Waimakariri Valley held in the names of Joseph Pearson and William Thomson and totalling 60,000 acres 'more or less'. These leases, issued by the Waste Lands Board, were dated 1859 and 1861, and no doubt were the result of Joe Pearson's exploration of the upper Waimakariri in 1857 when he made his pioneering trip on behalf of Joseph Hawdon, who took up Craigieburn, Grasmere and the Riversdale flats on his recommendation. The deed describes in elaborate detail the bank's security over the right of pasturage and, equally interesting, over the flock of 5000 ewes to be depastured thereon. It even contains a drawing of the registered brand by which these sheep were to be identified.

Picture my delighted amazement when I recognised an old branding iron, which I had found stuck in the wall of the brand-house at Grasmere when we bought the station in 1930. It had no relation, as far as I knew, to the Grasmere or Cora Lynn flocks, and I often wondered idly who the romantic character was who branded his sheep with a heart pierced by an arrow! Was it Moorhouse or Morten who had been stricken by Eros, and advertised the fact on 5000 walking monuments to his faithfulness? The answer to that we shall never know, but we still treasure that brand and I persuaded Mr Travers to have the deed photocopied and to lodge the original in the Canterbury Museum, where I hope the brand will join it.

The leases concerned appear to have been numbers 283, 284, 309 and 310, and to have comprised most of Lochinvar as it now is, though it is always difficult to disentangle the changes of the boundaries and owners of these early leases.

Chapter 9

THE CORRIDORS OF POWER

> Much have I seen and known; cities of men
> And manners, climates, councils, governments.
> > LORD TENNYSON, *Ulysses*

IT WAS CLEAR after the war that the high country was not going to prosper unless there was a substantial rise in the price of fine wools. The whole policy of the Labour Government was 'economic stabilisation' at the 1942 level so as to avoid the demands for higher wages which would cause inflation. They were right, of course, but you can do things in wartime that you cannot do in peace. However, the High Country Committee thought it was worth while approaching the Government on the grounds that the price of fine wools was still at an uneconomic level. The Wool Board had just been formed and we were lucky to have on it a staunch supporter from our own ranks, Harry Wardell, who joined us in 1944. We had had a meeting in August 1945 with the North and South Canterbury Catchment Boards at which they had agreed that pastoral occupation of the high country was still desirable and that it was in the interests of soil conservation that it should be profitable. Mr Machin, who chaired that meeting, was prepared to support us in our claim for a better price for wool for this reason and in the end we succeeded in getting an interview in Wellington with our own Minister, the Minister of Lands, Mr Skinner, and the Minister of Agriculture, Mr Roberts. Our deputation consisted of Bill Machin, to represent the catchment boards, Noel Jamieson, the Chairman of the Wool Board, Harry Wardell and myself. Meetings with Ministers are seldom very productive and this one was no exception. Many pious wishes were expressed and many noncommittal assurances given, but one classic remark has remained forever in my memory. In the past, Labour Governments have seldom considered it necessary to place agriculture portfolios in the hands of their most able men and Mr Roberts, who, as Minister of Agriculture, was responsible for the Wool Board, was what might best be described as 'rugged'. When he came to sum up his impressions of our discussion he began by saying: 'Now I'm not one of them people as keeps on belly-aching about the past. Wot 'as bin dun, 'as bin dun!'

In fact I heartily agreed with his sentiments in spite of their somewhat crude expression, and I hoped that he deplored the constant reiteration of the crimes of the early sheep farmers as much as I did.

Mr Skinner was a different type of man; he had served his time as a rabbiter in Central Otago, then had risen to become a Major in the New Zealand Army during the war. He was to prove capable of decisive action. At any rate he knew what the high country was like, which was more than one could say for most of the members of the Government.

Events were pressing thick and fast upon us during this period. Before we could get an answer to our submissions, the British Government called a conference in London to discuss the disposal of the huge stockpile of wool which had accumulated during the wartime 'commandeer' and the whole system of marketing thereafter. Australia, New Zealand and South Africa had sent delegations—ours led by Walter Nash—and until this issue was resolved it was obvious that we could not hope for any adjustment to the price of our fine wools, and that all our energies must be directed towards improving the administration of our lands and making satisfactory arrangements with the all-powerful catchment boards.

At the same time, and not unconnected with these international negotiations, there came a long overdue move to amalgamate the farmers' organizations throughout the country. There were two main ones: the Sheepowners'

Federation and the Farmers' Union—the 'squatters' and the 'cockies' in Australian parlance—and there was not much love and very little contact between the two. There was also a rather sketchy Crown Tenants' Association and our own infant High Country Advisory Committee, but these were of no real importance in the negotiations.

It was becoming very clear in dealing with a still-powerful Labour Government, with the dynamic figure of Walter Nash dominating the financial thinking of the party, that a united farming voice was absolutely essential, and New Zealand owes much to the statesmanlike men from both organisations who sank their differences and prejudices and joined to form Federated Farmers in 1946. If the Australians had done the same their farmers would have gained immeasurably in strength and influence and avoided some of the costly mistakes they made later on.

The immediate result to us of the formation of Federated Farmers was the opportunity it presented to put the High Country Advisory Committee on a firm basis and I began to negotiate with the emerging leaders of the new organisation. Outstanding among these was Gilbert Grigg, who was not only President of the Sheepowners' Federation, but also Chairman of the Meat Board. He had already been of inestimable help to us during our negotiations with the Government over an increase in the price of fine wools and our attempt to defend ourselves from over-zealous soil conservationists, and he was to continue to help us during our preparation of the high-country case to be presented to the Sheep-farming Commission appointed in 1947.

The members of our committee had paid their own expenses from the beginning in 1940 until 1945 when the Mackenzie Country runholders voluntarily subscribed two guineas ($4.20) each towards our expenses. This generous but well deserved offer led me to write to other high-country runholders and suggest they might like to make a similar contribution, and many of them did. However, it was clear that a stable and assured income was necessary if the committee was to function effectively, for it was most unlikely that the

Minister we were supposed to advise would dip his hand in the departmental pocket on our behalf.

My overtures to Federated Farmers were received with much more enthusiasm than I had any right to expect and I found myself almost embarrassingly welcomed to represent the high-country farmers on the committees being formed. We soon found ourselves at home in the Meat and Wool Section. At the same time, and much to my relief, the Federation accepted responsibility for our reasonable expenses. So successful was our organisation that the North Island hill-country farmers cast envious eyes upon it and formed a similar committee, for it was clear that, like ourselves, those who farmed the less productive country in the north had suffered severely from the fixed prices for meat and wool which prevailed during the war and as a result of the Government's stabilisation policy. This was so much the case that a demand arose for the appointment of a royal commission to examine the condition of farming throughout both islands, with special attention to the poorer hill country. Such a commission had been proposed and, I think, actually appointed in 1939 but had been abandoned when war broke out. We in the high country were not happy with its revival at this stage because it would obviously postpone any action which the two Ministers might take as a result of our representations on behalf of the South Island high-country farmers. However, as loyal members of Federated Farmers we could not oppose a proposal so obviously desired by the hill-country men, so I kept my mouth shut.

In due course the Royal Commission on the Sheep Farming Industry was appointed and the nomination of one of the members of our original committee, Willis Scaife, was accepted, which gave us great satisfaction, as we knew that his intimate knowledge of our country would ensure that its problems were understood. The chairman appointed to this august body was a well known farmer from Hawkes Bay, and I remember the truly Elizabethan language in which His Majesty appointed 'our trusty and well beloved Ronald Hugh White of Otane' to preside over its deliberations.

The preparation of the high-country case for the commission was obviously going to be a major task and it devolved almost entirely upon three or four already busy men.

It was to consist of:
1. A survey of high-country production estimating income from wool, and assessing stock, wages, interest, trade, rates, taxes etc. in order to demonstrate the immense value of the industry to the national economy.
2. A section on deterioration of country, stocking, burning and soil erosion.
3. A survey of land tenure and faulty administration.
4. An estimate of increased costs between 1939 and 1946, supported by properly prepared figures.
5. Aspects of taxation.
6. Housing, access, education and rural amenities.
7. Snow and kea losses.
8. Restoration of confidence in the future of the industry.

It was truly an ambitious project involving many meetings and long hours of work at home. The brunt of it was borne by Harry Wardell, Arthur Munro, Charlie Parker and myself, and I shall always remember the happy brotherhood of that period when we worked industriously together for a cause in which we steadfastly believed.

Harry and I had each been appointed members of a land board, as promised by Mr Langstone at our original meeting, and this gave us a close insight into the workings of the administration which we were pledged to criticise. A difficult situation, as it transpired.

I remained a member of the Canterbury Land Board until it was superseded by the Land Settlement Committees set up by the 1948 Land Act, and besides learning a great deal about land administration I learnt a lot about the departmental mind, which is often a strange and inexplicable one to the ordinary man in the street. I remember one occasion when we were discussing the financial problems of a returned serviceman who had been settled on a rather difficult property. The Field Inspector's report gave the man credit for his efforts to make the place payable and productive but made it clear that it would be a year or two before he could meet his obligations for rent, which was in arrears to a considerable extent. In such circumstances it was usual to postpone the current payments and to express a pious but rather doubtful hope that the arrears would be recoverable when and if the property became economic. After some sympathetic remarks from members of the board, our chairman, who was on that occasion the Deputy Commissioner, suddenly enquired what concessions the man had received from the State Advances Corporation over the payment of interest on his mortgage, and when he was informed that the S.A.C. had demanded—and got—its pound of flesh he cried indignantly, 'Well that finishes it; we're not going to make concessions to this man when the S.A.C. is getting paid!'

I was completely staggered at this attitude and pointed out that both Lands Department and S.A.C. were in charge of the taxpayers' money and it mattered not one jot to the taxpayer who made the concession. What we were concerned with was the care and preservation of the land in our charge and the soldier who had fought for it. We were NOT, in capital letters, a party to a struggle between two government departments!

The deliberations of the Land Board were not always as serious as might have been expected and one very vivid picture remains in my mind. This was a dispute between adjoining landholders as to who should have the tenure of a deep gully which was a piece of Crown leasehold. The rivals were two men of very different nature and, what was perhaps more significant, of very different build. 'A', who had leased the gully for some years and wished to retain it, was very tall, and lean and wiry, in spite of being well on in middle age. 'B', who had applied for the lease to be transferred to him, was short and stout—even a trifle breathless.

They were interviewed at different times, but strange to say, the stories they told were reasonably similar. 'A' said that the boundary fence which ran along the top of the gully and divided it from the property of 'B' was old and low and that 'B' deliberately stocked the adjoining land with cattle so that they would jump or smash the fence and gain the good feed and shelter in the gully, because his own farm was grossly overstocked. Nor would he mend the fence when the cattle smashed it down.

'B' said that the fence was not worth repairing and much the simplest solution was for our board to transfer the lease to him, whereupon there would be no further trouble!

What each in turn and in his own words described was how they had met at the disputed boundary one day when 'A' was chasing out his enemy's cattle, and how opprobrious names had flown between the two and how the lean athletic 'A' had chased the short and breathless 'B' across a kilometre of his own paddocks, flying the fences as they came—and no doubt thankful that they were no higher.

To sit in solemn conclave while such a lovely story unfolds itself from deadly serious and indignant lips is very difficult, and most of our members had a concealing hand across their mouths before the tale was done.

These regular meetings and all the others with which I was involved took me away from home a lot and left a heavy burden on Mary, who was left with a young family whom she was condemned to educate as well as feed and clothe. The Correspondence School is a fine institution but its administrators would never admit that the mother had anything to do except administer what they called 'a little disciplinary supervision'!

Famous last words! When Anne was in the stable feeding a horse and Ian and Robin half way up a tree and only the eldest, known as Tigger, helpful in the house; how was supervision —disciplinary or otherwise—to get school started at 9 a.m?

The High Country Committee started a campaign to get payments for mothers who were forced to teach their own children, because it concerned not only the big sheep farmers—who were expected to be able to afford help—but also a host of isolated working men's wives, who needed to be tempted to stay in isolation and tackle correspondence teaching. I made two separate visits to different Ministers of Education on this matter, and I remember well the impression created on my mind by Arnold Nordmeyer. I called upon him in his office late in the evening, but there was no sign of weariness in his manner. Alert, acute, active physically as well as mentally, he almost skipped about his well-carpeted room, like a bald, benevolent gnome, and I came away feeling that at least the points I made would be weighed in the balance, even if they were found wanting. Of course they were—no Minister can easily override the strong opposition of his departmental officers and this would have created a precedent of horrifying proportions.

It is interesting to compare this simple, and I think justifiable request, with the current opinion among feminist pressure groups, that all women who even stay at home and sweep floors are entitled to payment for the service they do for the community!

Our efforts had an interesting sequel; Mary prepared a paper criticising the Correspondence School for its attitude to this and other aspects of its work and by mistake inserted a copy in a returning set of children's work. She might just as well have put a letter bomb in the envelope; the effect on the school was immediate. A visiting teacher was despatched forthwith to interview this subversive lady who dared so openly to state her views. The school of course could not be expected to believe that the enclosure was accidental. The visitor arrived, ostensibly to examine the children's work and the background in their home, but he was obviously instructed to discuss the reasons for the criticism to which the Correspondence School was clearly unaccustomed. He turned out to be an elderly and serious gentleman for whom the lighter side of life had little appeal. The children found him heavy going and were none too co-operative about the work they displayed for him. That was no very serious matter but the evening, after school was over and the children safe abed, produced a minor confrontation which did little to improve our guest's opinion of us.

The evening meal revealed that he was not only a teetotaller but also a vegetarian, for whom the typical farm dinner of succulent roast mutton, potatoes, vegetables and rich brown gravy was not the most appropriate fare. This was disconcerting enough, but when we settled by the sitting-room fire afterwards, comfortably replete, at least as far as Mary and I were concerned, there came a crunch of gravel on the drive outside and the sound of manly voices, followed by a banging on the front door. When I went out I was greeted by a very cheerful party led by one of the most genial men I ever knew, Harry Nicholls. As Secretary of Quill Morris and Co., he had been my ally in the formation of the Bealey Hotel Company and acted as a director and secretary of that infant concern. Whenever he passed our way he always called in and sampled the bar, and on this occasion he had with him two representatives of a prominent Scottish whisky firm whom he had taken on a tour of the West Coast. Needless to say they were in a most joyful mood and, as they noisily removed their coats in the hall, Harry jostled me aside and handed me a bottle of their particular brand of usquebaugh, saying under his breath: 'Bring this out to offer them a drink and pretend you never drink anything else!'

I ushered them with some trepidation into our little sitting room and introduced them to Mary and our friend.

It was an uneasy mixture from the first, and when I went out and got a tray of glasses and the one and only bottle, our visitor's disapproval was as clear as day. When Harry, happily unaware of the deadly insult he was offering, handed him a well filled glass, he drew himself up with righteous indignation, and holding up a firm, rejecting hand he said: 'I have lived for sixty years without a drop of intoxicating liquor passing my lips and I do not intend to start now!'

Our friends were rather taken aback; they did not inhabit a world where people do not drink, and for a little while the convivial atmosphere was rather chilled, but the whisky soon restored their confidence and it was not too long before our more serious guest retired to bed.

The outcome of this visit was a rather sour comment from the Correspondence School on the children's work for the year—to the effect that they would make more progress if their supervisor was less critical of the school.

Chapter 10

DAWN OF A NEW LIGHT

And God said, Let there be light: and there was light.
Book of Genesis

OUR LIFE ON THE STATION struggled on through the icy forties, relieved only by the welcome and unexpected rise in the price of wool when the auctions began again in 1947. 1946 gave us the worst lambing for many years and this was followed by almost equally poor ones in 1947 and 1948. Bad weather in winter and spring seemed to be the main cause but in 1947 the Canterbury high country was struck by one of the freak summer storms which often make nonsense of the seasons. We were very late shearing that year because shearers were hard to get. We did not start until 23 January, and it was well into February when we finished. The weather was fine and hot as we shore the last of the ewes and dipped the dry sheep ready to take them out to Powers Country. Perhaps it was lucky we dipped them before they went out, as the cold bath is supposed to harden them up after the wool comes off. We drove them out to the high country on 10 February, and on 12 February my diary records the words: 'Fine and very hot.' On the 13th it says 'Fine and hot' but next morning there were 30 centimetres of snow piled up against the front of the house. How deep it must have been out on the tops where our newly shorn sheep were vainly cowering for shelter among the rocks we shall never know, but the storm cost us 1400 sheep and added that number to the shortage caused by the shocking lambing. Sheep will stand a great deal of cold even when freshly shorn but the change in temperature from over 30 degrees celsius to −15 degrees on the following day is more than any animal can stand—a range of 45 degrees in twenty-four hours. This is why the practice of pre-lamb shearing, so widely criticized by the uninitiated, seldom results in serious losses—the animals are inured to cold in early spring and the contrasts are much less when storms occur at that time of year.

That winter we began to carry out a long-dreamed-of plan: the installation of an electric power plant of our own, made possible by the rising price of wool. Ever since I had taken over Grasmere seventeen years before, we had depended on oil lamps and candles and a primitive and highly dangerous petrol light which served only the kitchen and one or two other rooms. Imagine a modern house bereft of electricity, not only for light but for all the multiple conveniences people enjoy today—washing machines, refrigerators, electric irons, vacuum cleaners, dishwashers, clothes dryers, electric heaters— even electric blankets—people think they can't live without them. Well, we had to! No wonder we dreamed of a glorious day when we too could share these wonderful amenities.

We had experienced one brief surge of hope in 1937 when the State Hydro Department put in a transmission line from Lake Coleridge to Hokitika to pacify the vociferous demands on the West Coast for a better power supply. Two powerful Ministers in the new Labour Government, Paddy Webb and Bob Semple, had close ties with the West Coast—which probably had a strong bearing on the decision. This 66,000 volt line was surveyed to pass right across the front of our house, looping its heavy cables right across our cherished view of Lake Grasmere. 'Visual pollution' as a catchcry had not been invented in those days but it was very real to us.

I went to see the engineer in charge and was amused by his puzzled surprise. 'Oh,' he said, 'most people are pleased to see electricity on their properties.'

'Maybe,' I replied, 'but most people hope to get some power from it. Are you going to offer us a breaking-down station on this line?'

'Oh well,' he said, changing his tone hastily, 'of course it would cost £10,000 to break it down for you and you'd have to pay for that!'

Ten thousand pounds ($20,000) was as far from our dreams as a million in 1936, so we resigned ourselves to sticking it out and hoping.

Before I went to Grasmere my predecessors had bought a small generator and petrol engine which was intended to provide low-voltage light in the house, but it was so small that it would only have served for seven or eight bulbs and those only in the house. I took one look at it and resolved that I would never put in a power plant until I could afford one which would serve all the buildings and give power for more than just light. Seventeen years was a long time to wait but at last the dawn of a new era was approaching.

I had consulted a very cheerful optimist about our possible water power resources, a man whom we all came to regard with great affection—tempered by a dash of scepticism about his estimates! He was Ted Salvesen, the technical half of a firm of electrical engineers, Wooff & Salvesen. Together we visited all the streams within a reasonable distance of the homestead because there seemed no point in using engines when nature provided abundant water power, if it could be harnessed cheaply enough.

There were three main possibilities: the old 'water race that runs uphill' which came out of the Cass Creek behind the homestead and drove a water wheel for crutching at the woolshed; a steep and rocky little stream which came out of the gulch which divided Mount Misery from Mount Horrible—gruesome names bestowed by weary musterers after some freezing day on the hill no doubt; and finally the placid outlet from Lake Grasmere itself which wound its way across almost flat ground and emptied into Lake Sarah across the railway line.

The first was too vulnerable to storms; every heavy nor'-west rain scoured the steep bed of the Cass Creek and washed out the intake or filled it with shingle. The second was a long way off, with the unbridged barrier of the Cass Creek to cross if anything went wrong. The third seemed likely to be the most stable and reliable, provided it could be diverted out of its normal course and dropped over a low terrace further down the outlet from Lake Sarah.

The high-country lakes are very various in their behaviour. There are some, like Lake Marymere, which have no apparent inlet and no outlet of any kind—true mountain tarns in which springs flow is really balanced by seepage and evaporation. Others are like Lake Pearson which is spring-fed but supplemented by the spasmodic flow of the Craigieburn Creek; so that its range of levels is enormous, sometimes providing a substantial overflow and sometimes none. Grasmere is a stable lake, fed by the spring drainage of the wide basin in which it lies, varying not more than 40 centimetres or so in depth and with a constant overflow. The plentiful spring water keeps it cold in summer, but in all but the severest winters ice is confined to the deeper water on the far side of the lake under the shadow of the bush on the Long Hill.

We spent a lot of time taking levels and measuring stream flows to determine the feasibility of the scheme and I couldn't help regretting that my father-in-law, Professor Alexander, died while we were at this work. With what boyish enthusiasm he would have thrown himself into the constructive work, for his favourite occupation when at Grasmere was to play with the intake of the 'water race that runs uphill'. In fact the water tables which we used in our calculations were 'Alexander's tables', compiled by an engineering brother of his, and standard works of reference for hydrological engineers.

After weeks of checking and rechecking levels we decided that it was possible to divert the Grasmere Stream across a flat piece of ground and that the fall into the Sarah Stream would be about 3 metres, depending on the height of the penstock we built at the top. Then came the problem of finding a suitable turbine to work with about twenty to thirty cusecs of water at such a low fall. Most private power schemes in the South Island use the type of turbine known as a Pelton Wheel, which is a high-speed machine utilising a small quantity of water at very high pressure directed into small buckets through a variable nozzle. Some people believe it is called a 'pelting' wheel for this reason. This requires only a small flow of water but a big fall—usually more than 100 metres. We could have used the system at the Misery site but the Sarah site required a proper low-pressure turbine. A Pelton Wheel can be made by any competent engineer but a low pressure turbine is a different proposition. Trust Ted Salvesen however; he discovered two turbines of an unusual type well suited to our special need. They had been installed many years before in a flour mill at Brookside on the Selwyn River; one to drive the main machinery and one to drive a lighting plant; it was the smaller one which he chose for us. The larger one was destined for a similar use and afterwards went to provide light and power for our friends the Ensors at Double Hill, up the Rakaia Gorge, where there was a bigger flow of water available. The turbine was only part of the installation, however. The war was only just over and materials of every kind were hard to find. A generator of a suitable size, ten kilowatts, was unprocurable so the resourceful Ted rewound a motor as an alternator. He then embarked on a project which must have cost him hours of time for very little financial reward. He built, out of bits of junk which littered Wooff & Salvesen's yard, an automatic hydraulic governor whose mysterious intestines are one of the great triumphs of a do-it-yourself engineer. Heath Robinson never designed anything quite so fascinating; its oil-filled stomach bubbles and boils like a witch's cauldron as a pump drives the oil through a series of pipes to open or close the turbine gate—all this at the touch of a control switch at the station, 3 kilometres away. If Ted had no other monument I would regard this as sufficient and there is little doubt that it will outlive its maker for it is going strong after nearly thirty years and he is an old, old man. (In fact he died before the publication of this book.)

Not only the power house, the water race and the concrete penstock had to be constructed, but the vital power line which would carry the current to the homestead, and this presented another set of problems. It had to cross the 66,000-volt transmission line referred to already, two roads, two telephone lines, and the West Coast railway line, and all these obstacles had different legal requirements. The railway, our ancient enemy, was the most difficult because beside the track was a line of poles which carried three sets of wires. On the top, almost 8 metres from the ground, were some high-voltage wires; below them on the same poles some lower-voltage wires for operating signals, and below them again were telephone wires. It was virtually impossible to go over the lot and so we had to get permission to dig a trench under the line and lay a ground cable to join the two poles on either side. It had to be 60 centimetres below the track and would have been a major task and we would have had to pay for railway staff to supervise while we dug it. I confess we cheated, and with a sledge hammer drove a 25-millimetre pipe through from one side to the other and drew our cable through it. It was not, I fear, 60 centimetres down where it came out but we covered it hastily and nobody has ever questioned where it went! We cheated again I regret to say at the main West Coast Road where the statutory requirement was 'stranded' wire, and we were using No. 8 fencing wire, which was all we could get. I found at the Ashburton Power Board some short lengths of stranded wire and wherever we had to make a joint we unlaid half the strands, wound their opposite numbers in and soldered the ends, rather like a long splice in a rope. As the joins were 6 metres in the air nobody ever spotted the deceit and they lasted for nearly thirty years until the whole line was replaced. The poles, after much searching of the West Coast for suitable silver pine, arrived at the beginning of winter. During the summer our good old tractor driver Tom Gibson had dug some forty holes, all 1.5 metres deep, to put them in and the heaps of earth beside the holes were damp with autumn rains. When we put the poles in during the winter we found the earth so frozen that it was impossible to ram it properly; when the thaw set in, accompanied by typical spring nor'-westers, all the poles between the road and the sheep yards took a lean of 45 degrees towards the east and had to be hauled up to a vertical position and stayed while the earth settled round them.

One curious piece of natural history emerged from this exercise. After standing open for several months almost every hole contained the corpse of an unfortunate hedgehog which had stumbled in under cover of darkness. I had no idea there were so many hedgehogs in the country although we had had a rather disgusting episode concering our house water supply some years before. We had noticed for some time a foetid smell in the bathroom and searched unavailingly for 'something rotten in the State of Denmark', but it was not until I drank a cup of water on a hot day that I realised it was the water itself which smelt so vile. The supply came in those days from a wooden tub fed by a pipe from the water race and when I went to investigate, there in all its loathsome putrescence was the body of a drowned hedgehog. Removing it hastily we congratulated ourselves on drinking clean and unscented water in future and thereupon forgot the matter; but to our dismay, in a few weeks' time the smell appeared again and we began to doubt whether the hedgehog had really been the cause. We were not long in doubt though because a visit to the tub revealed another of the same creatures drowned and disintegrating in our drinking water. After that a piece of wire-netting solved the problem, both for us and for the hedgehogs.

We were to have other lessons in natural history before the power scheme was completed; the major one concerned eels. Everyone knows that eels were one of the main sources of food for the Maoris and that they visited the high-country lakes to catch them at times. Most people have also heard how eels migrate to the sea—even, it is said, crossing dry land wet with dew by night to get there. Whether these tales are true or not it soon became clear to us that in the spring, if not at other times of the year, eels were going to pose a problem for our turbine. To begin with we did not install a screen to block them coming down the race and entering the down-pipe from the penstock, and soon after the plant was started the joyful celebration, which greeted the fall of a new night with blazing lights in every room, petered out in gloom and apprehension when the brilliance faded and died to a candle-like glimmer. Ted was with us for the great occasion and when next morning he and I went over to see what was wrong it didn't take us long to find the heads and tails of small eels blocking the curved blades of the 'runner' which turns by the force of the water coming through it. This was no joke because our turbine is what they call a 'casing' type in which the water is contained in a large cylindrical tank, from which it is drawn into the smaller cylinder containing the runner. To get at the runner you have to take a small plate off the side of the casing and insert what you can get of your body into its dark interior. Unfortunately we had not designed a very efficient method of shutting off the water at the penstock and, try as we would, we couldn't stop a considerable flow of ice-cold water pouring down into the casing from above. Add to that discomfort the stinking bodies of a number of slaughtered eels, some floating and some with heads firmly wedged in the louvres of the runner, and you have some idea of the job which faced us. We used knives and hastily-made hooks to cut and hack and haul the creatures out but it was a slow and difficult job. At one stage I went home to get some other instrument. When I came back I found an embarrassed Ted struggling into a pair of very wet trousers. He explained shamefacedly that he found working in wet clothes difficult and had stripped off completely and returned to his work inside the turbine. After a while he thought he heard voices and backed out hastily to find himself confronted by two girls, who had wandered up from the Canterbury College hut at Cass, and had obviously been regarding with some interest the sight of the lower half of a stark naked man sticking out of a hole in a tank!

After the eels were cleared we had to devise a method of keeping them out in future for it was obvious now that this was their period of migration. I made a wire-netting screen and dropped it into the penstock so that all the water had to pass through it and thought I had found the answer. So I had—for a little while!

The next crisis came when the shearers had just arrived and we were all at full stretch. During the night the lights went out and when I dashed over to the power house in the morning to see what had happened I found a scene of devastation. The slimy green algal weed which forms in lakes and streams in summer time had begun to drift out of the lake as the water got warmer and had completely blocked the screen; there was nowhere for the water to go except over the bank and down on to the power house below. The bank was collapsing rapidly and within an hour or two the whole power house might have been washed into the stream below. We got the water turned off in time to save any further damage and the shearers gave up their weekend leisure to fill sandbags and restore the bank, but it was a narrow squeak and I had to design another type of screen which allowed the water to flow over it in safety when it got blocked by weed.

There were other troubles, of course, the most dangerous of which was lightning. Thunderstorms are not as common in New Zealand as they are in England but in the mountains they can be monumental when they come, sometimes rumbling round the hills and going away and coming back for a couple of days on end. There was one I well remember, which struck the State Hydro transmission line in front of our house, completely decapitating two sets of poles, one on either side. The noise in the house was as if an enormous gun had been fired on the front lawn. This happened at about ten o'clock at night and Mary and I leapt up and rushed into the nursery where the two girls were sleeping. To our astonishment neither of them had as much as stirred.

If this kind of thunderbolt had hit our vulnerable power line very severe damage could have been done to the whole plant, so the wily Ted had designed what he called a 'horn-gap' lightning arrester, which consisted of two bent wires on top of the transformer poles at each end of the line with their points spread apart; at their closest point there was a gap of only 3 millimetres. One was connected to the line—the other to earth. In theory any lightning striking the line would jump the gap and go to earth. In practice two raindrops running down the wires at the same time caused a short which tripped the relay switch at the power house and shut the current off every time we had a heavy rain. Later on, when more sophisticated equipment became available, we were able to replace these primitive contrivances, and the trouble disappeared.

In recent years we have tried to persuade the power board which governs us to reticulate the whole area from Porters Pass to Arthurs Pass but each time they look at the prospective cost the guarantees which they demand leap from one impossible to another more astronomical figure and so there is little prospect of Ted Salvesen's monument being replaced until it rusts and rots away. What did it cost? Twice as much as he said it would. How much power did it produce? Half as much as he said it would. God bless him all the same; it's been a great success!

I don't think I blessed it more than on one wintry August night. Canterbury had been swept by a raging south-westerly snowstorm for three days and we had been to Christchurch for the races. Porters Pass was blocked by deep drifts; power was off all up the road and even the railway signals were out of action. We stopped at the Springfield railway station to see if there were any trains to get us home, and sure enough there was one at the station platform. Anxiously we asked if it was really going up the line. 'Well,' said the stationmaster grimly, 'it's going to start in that direction but God knows how far it will get!'

No light or power in Springfield, no signal along the track, it certainly seemed a forlorn expedition and engine crew and guard were not looking foward to getting bogged deep in snow half way and spending the night on their train. With the powerful headlight showing nothing but a shining white landscape in which the branches of the trees bowed to the ground with their burden of snow and only two faint white strips marked the line of the iron tracks, the engine drew out of the darkened station while we sat on hard wooden seats in the guard's tiny compartment peering out into the night.

The depth of snow was not so very great; the real danger was from drifts which might be piled across the track in isolated spots where the wind had driven it through some gap in the low hills. On we went, the staccato blasts from the funnel of the hard-stoked engine reverberating off the rocks in the cuttings, and dense black stinking smoke boiling alongside our little prison windows as we rumbled through the tunnels.

There are sixteen tunnels between Kowai Bush and Avoca—the seventeenth fell in while they were digging it long years ago—and all up the narrow gorge of the Waimakariri you dive into one black hole after another as the train winds and grinds protestingly around the bastions of the Torlesse Range.

When we finally rumbled over the Broken River viaduct and left the last of the tunnels behind, the view from our little window was of a deeply buried world. It was hard to tell how much snow there was, but at least the engine could still get through it and the long uphill drag began from Avoca towards the summit between Craigieburn and Cass. We stopped at Craigieburn so that the guard could phone Arthurs Pass to tell them where he was. There was a full 45 centimetres of snow here and it was sure to be deeper further up. There was nothing we could do but hope, and all the guard could tell us when he got back in was that it was not quite so deep at Arthurs Pass; what lay between was anybody's guess.

Painfully the toiling engine mounted the long slope, getting slower and slower until we were hardly moving. Near the top the wheels began to spin, the sudden release of power causing the slow gasping of the struggling engine to burst into a crescendo of rapid blasts. Then, as the driver throttled back, the wheels would bite again and the whole train inch forward a few more painful metres before the same thing happened again. Before we reached the top the driver was reduced to stopping the train and backing it down until all the couplings were loose, then plunging forward again so that each truck was started with a violent jerk. The loud metallic clangs ran down the train from front to back, each jerk more vicious than the last as the engine gathered speed, until they reached the guard's van at the back, which leapt wildly forward and almost flung us on the floor as we clutched at anything which might offer support.

The snow must have been 60 centimetres deep at the summit point and a few more centimetres would have stopped us there, but once over the top it was all downhill to Cass and the snow was thinning all the way. Gaily the train rattled down towards Lake Sarah and when we passed the low point between the Long Hill and Remus we looked eagerly across towards the south. Both together we cried aloud, 'The lights are on, hooray!' The house was visible from 3 kilometres away, and a little star-like gleam shone like a beacon in the waste of snow.

It was all very well to jump for joy but we still had to get from Cass to the bright warm haven of our house, and there were nearly 30 centimetres of snow in the railway yard. It would be a long and weary trudge if we had to stumble 5 kilometres through that! Luckily for us the foreman of the Ministry of Works was at home in his hut outside the station and he drove us home snug in his big truck. As we walked into the warm kitchen the lights were blazing bright and lighting all round the house the silver carpet which surrounded it. All Canterbury was in darkness, and here were we, comfortable, secure and independent. How we blessed Ted Salvesen that night.

The final chapter in the story of our power house was not written until the 1960s. We had always dreamed of the possibility of increasing its output which was scanty enough for the use of the whole station. House, cookshop, musterers' hut, stables, woolshed, cow-bail—all these needed lights and two of them water heaters and if any power could be spared we could always make use of it in household heating. Then there were the minor needs like washing machines, irons and vacuum cleaners, not to mention Percy's milking machine, which he had condemned as quite unnecessary but was only too glad to use! To get enough power for all these things out of a bare five kilowatts required a good deal of tightrope walking—the necessity for which we found the station cooks could never understand. There was a switch—turn a switch anywhere in New Zealand and you get all the power you want; why was it not so at Grasmere?

Knowing our need the faithful Ted Salvesen discovered a stack of iron pipes at the Christchurch Drainage Board. They had come from some old sluicing claim on the West Coast but when they were delivered to the Drainage Board yard the engineers condemned them as too old or otherwise unsuitable for their purposes. Ted was sure they could be bought for the proverbial song and used to extend the feed pipe to our turbine several hundred metres back upstream towards the lake. We did some amateur surveying and came to the conclusion that we could get an extra 1.5 to 2 metres of fall by going back until the intake was level with the water flowing under the railway culvert. Unfortunately Ted's calculations also showed that although the pipes were 45 centimetres in diameter we would have to use two of them side by side or else the friction of the flow of water against the walls of the pipe would reduce the gain in height to nil in power output. This meant that we could only go back less than half a kilometre because that was all the pipes there were.

It is hardly necessary to say that the 'cheap' price which Ted had promised us was quadrupled when the Drainage Board knew somebody wanted their white elephant and the final cost of about one hundred and twenty 6-metre pipes was a lot more than we had bargained for. However, once inspired with the idea we were not minded to turn back so we rather overhastily clinched the deal.

Then came the problem of getting the pipes home. A cartage contractor offered to pick them up from where they lay, cart them the 130 kilometres to Cass and land them on the job for about £90 ($180). Too good to be true of course. There was a railway line available and nothing on earth would make the Railways waive their right to cart the pipes—no matter what it cost the poor recipient. So we paid £60 ($120) to have them put on the rail; £90 to pay the rail freight; and then had the beastly things unloaded—by ourselves of course—in a vast untidy heap in the railway yard at Cass, from whence we had to cart them a few at a time, balanced precariously on our short-decked truck, over rough ground to where we wanted them. It's little wonder bureaucracy becomes a dirty word!

What with work and weather and economic ups and downs it was another three or four years before we actually completed the job, which was almost as big a one as the original installation. A line of old railway sleepers on which to lay the double line of pipes was the first requirement. Then tarred rope gaskets were made to seal the joints where the flanges of the pipes were bolted together. A method was devised to hold the pipes in position for bolting and a way of twisting them until the bolt holes were in line. The pipes of course were very heavy and I was rather proud of the simple bipod with which we did these tricks. Then there were traps because a few of the pipes proved to be of 50-centimetre diameter instead of 45 and, even worse, a few were more or less than 6 metres long which threw out all our calculations, and we could not construct anything at either end until we knew exactly where the end would be. No wonder it all took time.

At long last we were ready to construct the new penstock, 2 metres higher than the old one, and to build a new double embankment running back to the railway line. These we made three or four times wider than the original because we had learnt by bitter experience that if the banks were not wide enough they would develop leaks and require constant patching. Another lesson we had learnt was that efficient sluice gates were essential. It was impossible to do repairs—or extract eels—from inside the turbine with a deluge of ice-cold water pouring on your head.

When all was ready at the top end we had to design a box into which the two pipes discharged before the water plunged down the last steep dive into the turbine casing.

At last all was ready—or so we thought—but the long struggle was not quite over yet. First we found that the expansion and contraction of the empty pipes had loosened the joints and many of our sealing rings had dropped down inside the pipes and, most infuriating of all, someone had stolen two heavy rubber sleeves which we had had specially made to join the pipes at the penstock end to take care of this expected expansion and contraction, though once the pipes were full of water we knew it would be slight.

There was one final adjustment, though, which had to be made before we could test our great creation. The extra pressure of water would drive the turbine faster at its maximum efficiency and so the governor had to be adjusted to the new speed and the pulley on the generator changed or else it too would run faster, which it must not do.

DAWN OF A NEW LIGHT

At last the great day came. With bated breath we opened the two sluices and let the whole stream go surging down the pipes. In case some obstruction had been left in one of the pipes we had kept the end of the pipe open above the turbine so that the first flood of water could shoot straight out into the flax-filled stream below. As well for one poor creature that we did, because the first thing that shot out was an opossum; riding the crest of the great spout of water and clawing wildly at the air, it sailed in a long parabola and landed with a splash somewhere among the spears of flax below.

When we finally closed the pipe and let the water turn the wheels again there was an air of wonder and astonishment. The voltage climbed to 240 and stayed there without a waver; the generator hummed contentedly, and when we pressed the green button which throws the line-switch in, the governor took over its control job instantly, spinning the hand wheel over and opening the turbine gate to suit the load that was on the line. Better still, when we went home and put on all the load we dared we found it would carry 40 amps, nearly the 10 kilowatts that Ted had promised us back in 1948!

In spite of our jubilation over the result we were not to get off scot-free; there was bound to be a catch somewhere. It came in the form of an indignant protest from New Zealand Railways. We had raised the water level 15 whole centimetres beneath the tiny bridge that spanned the Grasmere Stream, without asking their permission, and we had constructed a bank on their property to make sure the water did not overflow! For this heinous offence they proposed to inflict the maximum penalty that their archaic law allowed and we were to be compelled to take out a licence and pay a fee to maintain the said embankment—all 60 centimetres of it—on their property. It was of no consequence to them that the bridge stood 2.5 metres above the water and that the 15 centimetres rise made no difference to its safety—we had done this dreadful thing without their permission and flouted the authority of a great public institution, a quite intolerable crime. In fact it was the result of my clumsy surveying!

Chapter 11

THE SHEEP-FARMING COMMISSION

> These are my politics: to change what we can; to better what we can; but still to bear in mind that man is but a devil weakly fettered by some generous beliefs and impositions; and for no word however sounding, and no cause however just and pious, to relax the stricture of these bonds.
> ROBERT LOUIS STEVENSON, *The Dynamiter: Epilogue of the Cigar Divan*

WHILE ROUTINE WORK went on at home; while the design and construction of the power plant were under way; while our four young children raced from their cradles to their school days; and while our little world moved wide-eyed and stumbling from war to peace, I was absorbed in the preparation of the high-country farmers' case to be presented to the Royal Commission on the Sheep Farming Industry. I worked in the evenings and on rare wet days; alone or with the other members of our committee; at public meetings where we sought information and statistics; and at private meetings where we collated and recorded them. Slowly we built up a case which dealt, in detail and with supporting evidence, with each of the eight aspects on which we had decided to make submissions.

There is little need to recall any of it except that part which constituted an indictment of the administration of the Lands Department. Ever since our first meeting at Tekapo in 1939 this theme had run through all our discussions and we were well aware that the department resented the criticism, but we were nonetheless determined to make our evidence on this point as powerful and as well documented as possible.

THE SHEEP-FARMING COMMISSION

The commission was to hear preliminary statements in Wellington and then to tour the country hearing submissions in each district, so we made out a general case for the Wellington hearing, to be backed by specific instances in each provincial centre in the South Island.

After the Wellington hearing, at which Harry Wardell and I presented the case, the commission set off on its tour of the North Island. When they were half way round, a bomb of almost atomic proportions fell from a cloudless sky; Jerry Skinner, the Minister of Lands, announced the presentation of a completely new Land Act to supersede the Act of 1924. The commission was enraged—the very issue upon which its report was to be partly based was changing under its feet. I was about to say the members were inarticulate with rage, but that was far from the case; they suspended sittings while the chairman, Mr White, and Richard Eddy, a retired union secretary whom Skinner had appointed to the commission, rushed off to Wellington to protest vehemently at this outrage. They got no change out of Skinner, however, and were forced to return to work and finish the task they had undertaken, in spite of the change in the rules.

The Director-General of Lands at this time was David Greig, a man of great ability and equal integrity, to whom the servicemen who returned to take up farming after the war owed the efficient and generous treatment which they undoubtedly received. There is no doubt that Dave Greig was alarmed by the strength of our criticism of his department, and the Act, for which he must have been largely responsible, was a wise and far-reaching piece of legislation designed to make fundamental changes in the whole approach to pastoral land administration, and to take into consideration the emergence of soil conservation as a factor in the Lands Department responsibility. It gave us a permanent right of renewal for pastoral leases with a thirty-three-year term, but it required tenants to accept stock limitations designed to protect the vegetation on erodible lands.

It was not only in the drafting of the Act that wisdom was displayed; but subsequently, in the organisation set up to administer it. A special branch of the Lands Department was created, with a Chief Pastoral Lands Officer and provision for four district P.L.O.'s who, independently of the district offices of the Lands and Survey Department, had access through their chief to the Land Settlement Board, which was the administrative body in Wellington.

Under this system there came into being just the sort of intelligent administration which we had asked for ever since 1939; and it was all the more valuable because the man who headed it was an ex-high-country manager, Ted Relph, whose dedication and determination made him a powerful advocate on our behalf. The system was rather less than popular with the Commissioners of Crown Lands, who had previously had sole authority in their districts, because they now had in their offices men who were directly responsible to someone else, and ultimately to a higher body; a situation nobody in authority enjoys.

Ted's approach to rent assessment—which was loosely defined in the Act as being required to be 'fair'—was to set rents as low as possible to encourage runholders to comply rigidly with the stock limitations imposed for the protection of the land. This attitude has persisted up to the present time, but once Ted was out of the way—he retired in 1969—the hungry wolves attacked his carefully nurtured team. The district P.L.O.'s were not replaced; there is now only a chief P.L.O., and normal administration is back in the hands of the Commissioners. Ted's conservatism in rent assessment has come under fire, partly from Treasury officials who think runholders should be made to pay higher rents, and partly from Lands Department officers who would like a rule of thumb by which rents could be calculated so that no knowledge or understanding of the land itself is necessary. So we are in danger of being driven back to the bad old days when we complained, with some justice, that the Lands and Survey Department was only a rent collector.

When the Royal Commission on Sheep Farming finally found its way to Christchurch I had the job of presenting the local case, including our criticism, supported by oral evidence of one or two instances of what we felt to be faulty administration by the Lands Department. I had warned the Commissioner of Crown Lands of the cases I proposed to quote because, as a member of the Land Board over which he presided, I did not wish to engender any personal antagonism; we were good friends and I hoped to remain so. When I had completed one of the cases and the ex-tenant of the run in question had given his evidence the Commissioner was invited to reply.

He did so by asking me whether I had not been a member of the Land Board when the alleged maladministration took place, and why I had not recorded my objection. When I replied that I did not join the board until some months afterwards he was completly at a loss to defend the case, having relied entirely on disposing of it by saying that I was a consenting party. It was rather a dramatic discomfiture and the administrators at the head office of the Lands Department were so upset that they persuaded the commission to accept evidence on the case at a special sitting in Wellington. However, I did not go and contest it on that occasion, having made a very effective point in Christchurch.

When the commission made its report we were well satisfied with the result—our submissions on a number of points had been well received; one of them was later the subject of legislation—the creation of a reserve for snow losses on the lines of the farm income equalisation scheme which eventually replaced it.

This legislation led to another trip to Wellington, one which threw rather an amusing light on politicians and their ways. It took place after the Government had changed from Labour to National and Sidney Holland was Prime Minister. I had asked for an interview with him to press for the introduction of the Snow Loss Reserve legislation and was told that he would receive a delegation of two at 9 a.m. on a Monday morning. As well say 3 a.m., because it was almost impossible in those days to reach Wellington from the South Island at such a time unless one was prepared to stay in a hotel and kick one's heels in idleness over the weekend. There was an early plane, but before the days of the modern Wellington airport it flew to Paraparaumu, whence it took the traveller as long to get into the city as it had to fly from Christchurch. Leo Chapman was to come with me and neither of us could afford half a week away from home. We therefore risked being late for the great man's appointment and took the early plane, trusting that there would be little of real importance on the Prime Minister's plate so early on a Monday. How right we were! The plane was late, of course, and we dared not wait for the official bus so we took a taxi all the way, threatening the driver with a complete breakdown of the New Zealand Government if he kept the Prime Minister waiting.

The greeting we received was friendly but terse: 'Now, gentlemen, I am a very busy man—how much of my time do you expect to take up?'

Bashfully I suggested that it would be difficult to explain the recommendations of the commission and discuss the means of implementing them in less than half an hour. At this he threw up his hands in horror and reiterated the innumerable pressures that devolved upon a man in his position. Without wasting time Leo and I got down to work and, once started, we found him helpful and sympathetic. After perhaps three-quarters of an hour his eyes strayed to a clock on the wall and I thought the guillotine was about to fall; but not a bit of it, he rang the bell and suggested that we gentlemen might like a cup of tea. Tea we had with buttered scones in a relaxed and genial atmosphere, during which he launched into reminiscences of his recent trip to England and his reception, not only by the awesome figures in the Mother of Parliaments, but by the Queen herself. An hour later we were still there listening to his description of the little drawing room at Buckingham Palace, where he had been entertained, as we were being entertained, with a friendly cup of tea. When we finally left—without, I might say, any interruptions suggesting an intolerable pressure of business on his shoulders—it was with the assurance that he would get his law draughtsmen to look at the proposals and try to frame legislation to suit the case.

The law was eventually passed as part of a finance bill; it

THE SHEEP-FARMING COMMISSION

provided that a farmer in country liable to severe snowstorms might put into a Snow Loss Reserve such part of his income as he thought fit and draw it out to supplement his income after a loss. Who could partake in this scheme? Well obviously if you did not see a likelihood of losses you wouldn't put money into it—so there was little need to define participants. How would you establish your losses? That certainly seemed a problem, for the Commissioner of Taxes could hardly go out and count dead sheep. It was left to the farmer to make a statutory declaration of losses, which could include anything attributable to snow, even lowered wool yields and poor lambing percentages.

The whole thing impressed me with the innate fair-mindedness of the New Zealand Department of Inland Revenue, a characteristic perhaps few taxpayers appreciate. When I came to serve on a Snow Loss Reserve Committee, presided over by the District Commissioner of Taxes, I was struck by his easy-going attitude towards the administration of the scheme. I had some reservations at times about our decisions because I often had information unknown to the other members of the committee. Not that the claims were false—many high-country stations have some snow-caused losses every year—but I often knew the real reason why the money was needed urgently. On one occasion a daughter was getting married and a large wedding was contemplated; on another a new house was being built. The Snow Loss Reserve was a handy source of available cash, and as the Commissioner remarked: 'It's his own money and he pays tax when he withdraws it.' In the light of this kindly and tolerant administration it was a pity that some accountants advised their clients not to use it for fear it might be difficult to get the money back; in fact it was surprisingly easy.

The story of our trip to Wellington presents the late Sir Sidney Holland in rather a trivial light; actually he was anything but trivial, and I always remember with admiration his handling of the Wool Retention Scheme in 1950. The wind-up of the Joint Organisation which linked Britain, Australia, New Zealand and South Africa in a partnership to hold and market wool during the war was ably handled by Walter Nash and the newly formed Wool Board on New Zealand's behalf; and with our share of the profits which accrued stowed away in a reserve fund for the support of wool prices should they fall too far, we were able to return to the auction system of wool selling. To the great delight of high-country woolgrowers, not only did world prices rise rapidly for all wools, but for our merinos and fine halfbreds the demand for fine fabrics after years of austerity drove their relative value to a handsome premium over the coarser wools. Harry Wardell's great cry of 'relativity' was vindicated by the world markets, and the high country faced the greatest period of prosperity it had ever seen.

It was not before time. The long years of neglected maintenance and development had taken a heavy toll of our properties, and before any of this bountiful reward went into our pockets in the form of luxuries, there was much to be spent on buildings and fences and manures and seeds. It is to the credit of the high-country farmers as a whole that this is where most of it went. Some, like ourselves, lightened our burdens with power plants to make life more tolerable for our families; others, also like ourselves, grabbed at the bonus with heart-felt thanks to pay boarding schools' fees to educate our children; but generally the first thought was for investment in the productivity of our land; for even after all the years of struggle and low profits, hope and courage had never deserted us.

The boom lasted for several years and culminated in the Korean war of 1950, which set off a frantic demand for wool, led by the Americans. When the selling season started in August 1950 it soon became obvious that prices were going to rise to heights beyond our wildest dreams and that money was going to flow into New Zealand and Australia in embarrassing quantities. This was when Sid Holland took a hand. Whether the scheme was his idea or that of his Treasury advisers I do not know, but he called the leaders of all branches of Federated Farmers to Wellington to meet him and hear his proposals for a freeze of part of our wool receipts.

I well remember the scene as we assembled in the vestibule

of a large room in Parliament Buildings. Word had got round before the meeting as to what he was going to demand, and the air was full of dire threats from delegates from country districts where hard physical work was more common than secondary education. 'He's not going to get away with this,' they cried. 'It's our money and we've a right to spend it if we want to. Why should we be singled out for control when everyone else can spend what they like?'

The mood was definitely hostile and when the Prime Minister arrived and greeted us he must have looked round a circle of grim faces and gained little encouragement from what he saw. However, he went ahead and calmly but forcefully explained his reasons for asking farmers to put into tax-free and interest-free reserves one-third of their wool cheque in the current year. He explained the inflationary effect of large and unexpected sums of money appearing on the economic scene when there were not the goods and services to spend them on and he wound up with a telling point when he said: 'I'm not asking you alone to bear this burden. If this money is received by you more than half of it will find its way into my hands through taxation. I don't want you to have it and I don't want it either; if I have it every pressure group in the country will be on to me to spend it!'

After his speech he announced that he would leave us to talk it over and come back when we sent a message to say that we were ready. Then the fun began. Those who had come with their minds already closed against the scheme were vociferous and appeared at first to dominate the meeting; but after a while a few of the wiser and more experienced leaders began to speak and to show that they could see a great advantage to ourselves by being allowed to equalise our incomes by drawing this money when we needed it more. Tax rates were very high after the war, and in the majority of cases only one-third of this money would have remained in the woolgrowers' hands.

After much discussion a compromise was reached in which it was decided to accept a freeze of twenty-five per cent, and the Prime Minister was summoned to hear the verdict of our great democracy. He took it calmly and thanked us politely for our cooperation, but then, having lulled our suspicions with soft words, he suddenly turned the tables on us with consummate skill.

'I think,' he said, 'that you have made one mistake. You have confused me with my predecessor [Walter Nash, though he did not name him]. If he had needed one-third he would have asked for one-half and hoped to get the lesser amount. I do not treat people like that. If I needed only one-quarter I would have asked for only one-quarter knowing that you would trust me as I trust you. I need the full one-third to protect New Zealand against the effect of this sudden surge in income, and I trust you honest farmers to let me have it.'

So he went away again, asking us to reconsider our decision. This psychological medicine worked to perfection, for he got his one-third with only a few dissentient growls.

The principle was established, but we then had to work out an administrative plan acceptable to us and the Government, and, together with such august figures as the Secretary of the Treasury as well as my well known colleagues at the head of the Federated Farmers and the Wool Board, I found myself on the committee set up to examine ways and means. Before our first meeting I took the opportunity of attending meetings of woolgrowers in Canterbury to find out what they would be prepared to accept. Canterbury is a great wool province and it has always had a body of far-sighted men with an understanding of finance as well as a dedication to wool—mostly fine wool at that. At one of these meetings in North Canterbury a well known Christchurch accountant, Bob Free, who had many big woolgrowers among his clients, made the suggestion that the frozen money should be paid back year by year over a period of ten years, to spread the income evenly and let the farmers know exactly what they would get each year. The scheme impressed me, and after discussion over safeguards and provision for death and other unexpected disasters, I decided to take it to the Wool Retention Committee as a concrete proposal. In the end it had to be accepted in a modified form because the leaders of Federated Farmers did not think their critics—some of whom had not forgiven our capitulation over the scheme as a whole—would accept more than five years.

As things turned out we would have been wise to accept the ten-year period, for the price of wool remained fairly high throughout the 1950s, and inflation had hardly begun to take a toll of frozen assets at this time.

There was an amusing epilogue to this affair. One of the most outspoken critics of the scheme was a prominent Canterbury farmers' leader, a man of substance and considerable influence. Towards the end of that season I met him in the street.

'I've just come from my accountants,' he said, grinning all over his face.

'I've frozen every penny I could get my hands on and it's going to save me a packet in tax!'

At least his accountants had seen the benefit, if he had not.

Chapter 12

THOSE BLOODY BIRDS AGAIN

> Wild animals never kill for sport. Man is the only one to whom the torture and death of his fellow creatures is amusing in itself.
>
> JAMES ANTHONY FROUDE, *Oceana*

NOT ONLY were the 1940s a period of bad weather in the high country, reducing our profits by snow losses and poor lambings, but in our area at least we suffered severe attacks by keas. When we were mustering for shearing at the beginning of 1949 I saw, for the first and last time, a flock of keas actually attacking sheep. We had mustered the Burnt Face at Cora Lynn where our older wethers had spent the winter, and besides finding several of the sheep in the mob with open wounds on their backs, the men saw isolated sheep dead on the hill. When sheep have died of starvation after being trapped in the snow they are usually in small groups left behind when the snow melts. When you find them in singles it is often because they have been chased away from their mates by a determined killer and died where they lay exhausted after a long chase.

We made several trips back to the Burnt Face to look for the keas, but saw nothing until one day two of us went up the Bealey Spur which lies across the precipitous gulch of the Bruce Creek and from which the whole of the great face which comprises the slopes of Mount Bruce and Mid-Hill is spread out before you. There is no doubt that the Bruce is a favourite nesting ground for keas, but it is so rugged and inaccessible that the finding of their nests would be a difficult job. Their natural diet of berries, grubs and leaves is plentiful enough except when the ground is covered by snow, but once a kea has developed a taste for meat he will continue attacking sheep until he dies, and on the Burnt Face in winter he had an easy target, for the big wethers were often trapped in snow for some time.

As we looked across the valley from the Bealey Spur that day our attention was caught by the birds' tell-tale cry. It is not often that they attack in broad daylight—usually at dawn or dusk—but there must have been a few stragglers left on the block after our muster and they were too tempting for the killer. We were a long way off, it is true, and I cannot say that I saw a wound inflicted or a sheep die, but through the field glasses it was easy to see a bird swoop down on a sheep and see it plunge away downhill, followed by a dozen screaming tormentors until it disappeared in the bluffs which fell into the river and was lost to our sight. It was impossible to climb down there and doubtful if we should have found it if we had, and anyway we were looking for stragglers on the block we were on and didn't have the time; but the killer was still at large the following August when we shot seven keas after another bout of killing. Did we get the killer then? Who knows?

The autumn muster that year was bedevilled by a lot of bad weather. We went up to the Top Hut and got the first day in but we were a man short because a casual musterer whom I had engaged couldn't come when I wanted him. He was to arrive the next day, and as he had never been on the place before, somebody had to go down and bring him up to the hut. As it turned out it was pouring with rain so I went down myself. For some reason we did not start up the bush track until after dark and in the pitch black night even our two horses, familiar with the track, kept wandering off it into the forest. For the first few kilometres, where it is continuously among the trees and where there are many unexpected turns and twists, I was reduced from time to time to getting off my horse and searching for hoof marks by the sputtering light of a match, often extinguished by the wind and rain as soon as it was lit. My companion, George Pilcher, whom I had not met before, proved to be an amusing character with a fund of stories picked up in years as a casual musterer, deershooter and, as he described himself, 'general bush-bastard'. He told me how, as he came up in the train that day, he was on the wrong side to get a view of one of the station homesteads he had often heard about. He couldn't bear to pass it by without a closer look, but when he crossed the aisle to gaze with rapt attention at the sheep yards, woolshed, dog kennels, etc., which were to him a fascinating and absorbing sight, he found his view obscured by a pert young lady, who perhaps thought that she was the object of his interest. Far from it—dog kennels and stables were of much greater importance to him, and he remarked to me: 'I don't know what the flash sheila thought of this bush-bastard peering over her shoulder but I'd heard so much about the place I just had to have a look!'

Apropos of our own situation, he told me about one very young musterer who had little eye for country and was constantly getting lost on the hill. 'I said to him,' said George (after one frustrating search party) ' "three quid a week and found" doesn't mean found every time you bloody well get lost!'

So while we struggled through the bush and finally emerged into the open pakihis of the Bealey Spur he beguiled our way with anecdotes. Here the rain eased and the racing clouds swept across the sky from the summits of the Shaler Range and fitful moonlight lit the way for us so that we completed the rest of our midnight ride without losing our way or getting bogged in the swampy lagoons which lay across the path.

It was still raining the next day, and when we finally began the muster again sheep had gone back to the Bruce Saddle from where they had been mustered on the first day, so we had all that to do again. That was not the end of a very troublesome muster either, for on the third and last day we must have bypassed a large mob of sheep which had run down the Jordan Spur into the bush, and some days later one of the Government deershooters saw them after they had climbed back into the basin above. There was nothing for it but to go back and try to get them, so I decided to take George Pilcher and walk up to where the Jordan Stream emerges from its deep bush gully on the Top Flats and up the Jordan Spur which debouches into this gully a kilometre up from its mouth.

In late autumn the days are short and after we had walked the 5 kilometres from the Waimakariri bridge the sun was well on its way before we started our climb. It was longer and steeper than I had expected and the track bore unmistakable signs that not only deer used it regularly but sheep also, and I began to suspect that this was one of the ways by which they came down from Powers Country, which we had long known they did from time to time. The climb through the bush was wearying enough, but when we emerged into the open country at the top of the Little Jordan Spur, the basin, we soon saw, was full of sheep; this meant that we must climb swiftly and silently and as much out of sight as possible in case we started them stringing up towards the saddle at the top before we could get a heading dog around them. Once in the saddle there were 300 metres of open slope down to the main Jordan Saddle and if they got away down there, there was not enough daylight left to head them and bring them back again.

Without a pause for a breather we clambered up the knife-edged ridge which fell away in cruel-looking bluffs on the west side, forcing us to climb in full view of many of the sheep, until at last we came out on to gentler slopes and sent a dog to save us the last 150 metres. It was a very worthwhile expedition, for once the sheep were turned downhill towards the Hut spur we could get a rough count as they ran along the ridge and see that there were not far short of two hundred. Once past the hut we could let them go to find their own way down the Bealey Spur; but George and I still had a long and steep walk down to where our truck stood at the bridge.

1949 saw the first of what was to become a series of annual pilgrimages to this same hut—not for the sake of sheep, but for the family. Above the hut the spur rose about 150 metres to the little peak that was really the end of a long, knife-edged ridge which swept down from the top of the Black Range and formed the high-road up and down which we drove the sheep. The slopes of this peak were a delightful series of undulations, some steep, some easy grade, which made a perfect learners' slope for skiing. In the September holidays we packed skis and sleeping bags and the younger children—Robin was only six and Ian eight—on hacks and packhorses and set off to camp and ski. There is nothing in family life more rewarding for parents than an expedition in which the whole family takes part, and when it takes place in wild and unspoilt country with snow-covered mountains as a background and the dark green of the bush with its floor of sifted silver all around, the whole proceeding takes on a glamour and excitement like some dream journey into an imaginary land. How the children laughed and chattered as we loaded the horses, balancing tucker boxes and bedding rolls and their short skis and sticks, so crude compared to the expensive equipment of today. How proudly they strode out up the track to start with and how tired their little legs got before we were half way there! But when we reached the hut the excitement began again; the choosing of the bunks; the gathering of wood; and the first lovely scent of birch smoke as the fire began to blaze up the battered iron chimney. Then, when night fell and the glittering candles lit the tiny room, how warmly and cosily we clustered round the fire and giggled at the events of the journey up. And when we had fed the hungry mouths and tucked their owners into sleeping bags a book came out to read them off to sleep dreaming of some imaginary land of make-believe.

Next day we set off to explore the ski ground. I said that our equipment was crude by modern standards; in fact, to make it possible for the children to climb the hill we contrived cords criss-crossing the bottom of their skis so they did not slip backwards. Even the six-year-old had been broken into skis on the flat paddocks at Grasmere almost since he could walk, so getting up the clearing and through the bush was not beyond his capacity; once on the open slopes above they could take off their ropes and flounder and fall and roll and laugh until they were tired and hungry.

It was all so different from the introduction to skiing that children get today on smooth and gentle snow slopes with rope tows and lifts to take them up the hill and access in a comfortable car to the place where they can don their skis.

We mapped a path down through the strip of bush, and over the years its features got named by things that happened there and by places in the books which we were reading at the time. There was the Shambles were everybody always fell and got inextricably tangled in stunted shrubs; the Deer-Wallow whose muddy bottom was apt to melt the snow into a slushy hole; the Scenic Railway where the humps threw the unwary into the air; and the long curved run where you emerged from the bush was Regent Street. Later on, the place of departure at the top was named Campo 12 when we were reading *Farewell Campo 12*, the escape story of two New Zealand brigadiers.

The first day's frolic ended early, from sheer weariness. Our return to the hut when the older ones came skimming down the gentle sunlit slope gave rise to a saying that is with the family still. Little, snow-plastered, six-year-old Robin came staggering and floundering behind the others to collapse in a heap at the hut door, crying, 'Here comes what's left!'

Unfortunately, when we went to feed our horses that evening we discovered that they had revolted against imprisonment in their snow-covered yard, pushed the flimsy netting gate down and departed. They were by then no doubt kicking their heels outside the stables at Cora Lynn. We stayed a couple of days until one of the men found them and brought them back and then we all went home.

We made many trips up there, until the children's advancing skill eventually drew them to more sophisticated ski grounds with rope tows to pull them up the hill and spacious huts to sleep in.

If camping in mountain huts and skiing developed the children's initiative and competence in winter, summer brought them even more participation in outdoor life and the work of the station. Much of it was connected with horses, and though they all rode the station hacks when we were going anywhere, Anne was the one truly born to the saddle. I remember her at the age of three, standing in a ring of animals four times her height; they snuffed suspiciously at her confident little figure while she held out her hand in welcome saying, 'You know me—I'm Anne!' She was ten years old now and we wanted to send a mare in to a stallion they had at Mount White. It was a long trip by the road and I had no man to spare to take her in, so we decided Anne could ride the mare over and come back in the Mount White truck when it came out for the mail. To save time she was to ride across the Waimakariri River instead of going the long way round by the bridge. Even the shortened route would be 25 kilometres. It was the old way known as the 'bullock track', used by the wool waggons before the river was bridged in the 1870s; still an almost navigable road on our side of the river but hard to follow on the other side, where it climbed a series of terraces left by changes in the river level after the last ice age. (This area offers a fascinating study in geology because in certain lights you can discern eighteen separate and distinct terraces which must have marked the changing levels of the river as it cut down through the rocky gorge which divides the upper valley from the plains.)

The whole family rode down to the river to see the child safely across and make sure she chose a safe ford. The river bed is wide here, which is why they chose this place for the waggons to cross, but the channels change and the water can be very swift. The mare, Sally, was good in the water but she was only fifteen hands and small enough if it was deep. We found Anne a ford through the biggest stream, starting her high up at its widest point so that she could ride diagonally down towards where it began to narrow again, so getting the benefit of the shallow water and not having to fight against the current. She waved cheerfully to us and plunged into the cold clear water in which rocks and stones are magnified and their colours brightened by its translucence. We watched Sally's legs going deeper and deeper until they disappeared and her belly touched the surface. The child's legs were short enough not to get wet as long as it got no deeper and we watched anxiously as the little figure on its half-submerged mount got smaller and smaller. We all heaved a sigh of relief when the horse's legs began to appear again as the water shallowed on the far side.

There was little to worry about after that. I had warned her against soft sandy patches where she might sink in. The Waimakariri is not bad for quicksands, though the Rakaia and the Rangitata can be very dangerous if you are not careful where you ride. We saw her gain the solid ground on the other side of all the streams and set off down the wide flat called Poverty Flat, a hungry and once rabbit-infested piece of ground left by the wide curve of the river which always runs on our side at this point.

A tiny hand waved us farewell and we left her to scramble up the terraces and ride the 16 kilometres to Mount White along the road.

It was not till evening when Helen Bell, the manager's wife, rang up to say she had got there safely that we heard of her adventures on the way, and not till next day when she came home that we got the detailed story. When she got up near the road she suddenly saw a pair of Canada geese with a brood of well-grown goslings. We never let a chance like that slip and, true to her upbringing, Anne set off after them on Sally. Young geese will run for a while and then flop down at full length among the tussocks hoping you will not see them—the old ostrich trick. Seeing one hiding, Anne leapt off her horse and grabbed it by the neck and, as she had seen others do, she whirled it round and round until the neck was twisted and the bird hung lifeless in her hand. She had lost track of the rest of the geese while this was going on and Sally was showing signs of walking off and leaving her marooned on foot, so she hastily recovered her means of transport, well satisfied with what she had. There was a cord on the saddle so she tied the bird's neck firmly to the pummel and, climbing aboard again, she set off along the road. Unfortunately her execution hadn't been entirely successful and before they had gone very far the creature began to come to life again, beating its little wings against the horse's shoulder and kicking wildly with its legs. Sally took a poor view of this demonstration and Anne had to leap off hurriedly, untie the flapping bird and repeat the treatment until she was quite sure it was dead. She reached Mount White and stabled her horse and Helen was very amused to find her small figure standing at the back door proffering with one hand a dangling flapper goose.

Chapter 13

WATER, WATER EVERYWHERE

> Then banks came down with ruin and rout,
> Then beaten foam flew round about,
> Then all the mighty floods were out.
> JEAN INGELOW, *The High Tide on the Coast of Lincolnshire, 1571*

IN THEIR CAMPAIGN against the burning of tussock country the catchment boards put a great deal of pressure on us in this post-war period. In fact most of the high-country men had ceased to use burning except in very special cases where the vegetation, which usually included some sort of scrub—manuka, matagouri, cassinia etc—had got so rank that it was an impediment to the movement of stock. We frequently warned the boards, however, of the danger of accidental fires occurring when large areas of the country became equally rank. The nor'-west winds which plague the Canterbury gorges can sweep a fire for miles and undo the work of a generation in an hour or two. The Waimakariri Valley was particularly vulnerable because of the railway. In those days of steam locomotives there seldom passed a spring without several dangerous outbreaks, lit by engine sparks and spreading like the proverbial wildfire through the stalks of grass beaten into a dry carpet by the snows of winter.

I had many discussions with the soil conservators about the need for protective burning and for many years the railway staff were supposed to burn the borders of the track to prevent fire spreading on to the adjoining hills; this was only partly effective as the sparks often flew beyond the burnt strip. The Chief Soil Conservator of the North Canterbury Catchment

Board, R. D. Dick, believed that it would be possible to burn firebreaks round the bottom of the more vulnerable hills, provided there were enough men present and it was done carefully. I didn't agree; I knew the wicked gusts of wind which come from nowhere when a fire begins in still weather and I knew how pitifully few the still days were in spring. You might gather a gang of men together a dozen times and each time, just when you were ready to start your burn, the clouds would begin to clog the passes in the west and the first warning puffs would cool your cheeks. Then they would all have to be sent home to try another day.

In September 1949 an experiment was tried round the foot of Mount St Bernard on the boundary between Grasmere and Craigieburn. A dozen or so students from Lincoln College had been assembled for the job and the day looked promising enough. Fortunately I had no part in it and was not even at home for most of the day. Towards evening a gang of blackened and exhausted warriors came in to Grasmere for explanations and refreshment. They had burnt a strip 400 metres long and perhaps 20 wide along the boundary fence and in the process had lost skin and hair and sweat enough to last them many days. The foot of Bernard must be at least 20 kilometres round. How many years would be required to complete the firebreak? It is not surprising that it was never tried again.

A more significant experiment began a little later on, however—one which although it failed on that occasion, pointed the way towards a revolution in the high country. Aerial topdressing had already started in the North Island with promising results, but nobody seriously believed that the sparse unproductive pastures of the high country could be improved significantly by a dusting of superphosphate from the air. It might be possible, however, to improve the swards by sowing seed which might germinate in the bare patches amongst the tussocks. The same Chief Soil Conservator sponsored the first experiment in the district and, thinking that native species might be the most likely to germinate in these conditions, he chose *Notodanthonia pilosa* for the trial. There were no aeroplanes designed for this sort of thing and in fact the only aeroplane available was an old De Haviland Moth; a primitive small biplane with an open cockpit. The trial took place at Craigieburn where there was a good smooth slope for a take-off ground and plenty of low tussock hills near by to scatter seed on. There was one unexpected snag though. Danthonia is a seed with a hairy awn which makes it cling together, and as the seed is very light you get a ball of fluff rather like a dandelion head. The plane took off and flew around but when the trapdoor was opened to let the seed flow out it remained exactly where it was. The poor pilot tried innumerable tricks to shake it out, flying along and letting his little plane down in a series of unnerving bumps, but it was all to no avail—the fluffy seed just would not come out. As a demonstration it was a failure—as a milestone it was significant. Later on scientists treated danthonia seed to knock the awns off in an attempt to make it flow, but by that time its already low germination rate had fallen even further and I doubt if any of it was ever actually found growing.

1950 and 1951 were notable for two floods which caused major problems to us, disrupting transport services and soil conservation efforts. The two floods were so devastating and so dissimilar that it is worth recording them.

The 1950 flood took place in May and was the result of torrential nor'-west rains along the main divide. Its effect was not only to disrupt the road and railway in the upper gorge but also to enhance the constant threat which the Waimakariri poses to Christchurch City, a threat which has been since 1868 the great preoccupation of river administration. The main ranges of the Southern Alps have been little affected by the actions of sheep farmers. Generally speaking they are still well covered with native forest, mainly mountain beech, and the only major modification to their plant cover has been from the introduction of what are known either as 'game' or as 'noxious' animals according to your personal prejudices. If you are a 'sportsman' they are game; if you are a soil conservationist they are noxious animals. Whatever name you call them, deer have been the chief destructive influence in the headwaters of the major rivers, beyond the more open sheep country. It was from these areas that the 1950 flood mainly came, turning the Waimak into a raging torrent, washing away a span of the railway bridge at Cora Lynn and some of the track below the bridge, and cutting into the West Coast Road as well. Nor'-west rains are the predominant source of rainfall in the high country and the rainfall figures follow a fairly regular pattern—the annual 130 centimetres at Grasmere become 180-odd at Cora Lynn and 400 at Arthurs Pass. This progression is often accelerated in really big storms so that a fall of 13 centimetres at Grasmere—not unusual over a couple of days—becomes 40 to 65 at Arthurs Pass. It is no wonder that the saturated and depleted forest cover cannot contain it and it thunders out of every small stream, carrying debris and fallen trees out into the main river where its swirling torrent washes great banks of shingle from side to side and down towards the gorge, accompanied by the dull thud of boulders rolling in its bed.

We drove up the road in the truck to see the damage when we heard the bridge had gone and walked across its narrow planks to see the broken span on the far side. The children were much amused by the antics of the railway men who were trying to pass a rope across the gap. The line was actually still complete, its rails with the sleepers attached sagging across the missing span in a deep and dangerous-looking loop. One man took the end of the rope and crawled precariously across this flimsy bridge, but the children's admiration was reserved for another who boldly stood upright and stepped across from sleeper to sleeper like Blondin across Niagara Falls, while the flood still roared and rumbled 6 metres below him.

We were absorbed in watching this exciting scene when suddenly I looked back and saw behind us the line of tall poles which carried the many power and telephone lines across the bridge beginning to collapse. They stood on the upstream side of the bridge and no doubt their footing in the river bed was being washed out slowly by the surging stream. I had little doubt that all the current had been cut off but it might just be that it had not, so I gave a panic cry to the children and we all raced back across the narrow plank as the whole line slowly subsided over where we had been standing, completely blocking the path by which we had retreated.

A few days later we went back again to watch two bulldozers which were to try to shift the river away from our side so that the track could be repaired where it was swinging in the air down stream from the bridge. The children were enthralled to watch the two giant machines—smaller and a far less common sight than they are today—linked by a wire rope in case one of them got into deep water and testing the depth as they proceeded by dropping the leader's blade from time to time. The leading machine was driven by a fair-haired giant who looked like a knight in shining armour riding a great monster of a horse into battle, and for a flag upon his lance he flew from the flagstaff of his exhaust pipe a pair of underpants which had got wet entering the water. We heard that he had been a famous driver during the war, working fearless and bareheaded making airstrips in the islands of the Pacific while the raiding Japanese fighters dived upon him from the sky, trying to destroy each strip before it could be used. The powerful river was too much even for these big machines however and, until it fell to its frozen winter level, there was little that anyone could do to shift the banks of shingle.

The flood which came the following year was of a very different kind. This storm also was an autumn one, this time in the middle of April, and coming from the east. It struck the front ranges and did not penetrate very far towards the main divide, depositing its destructive deluge on the depleted ranges where 100 years of fire and grazing had had their greatest effect. What would have been the result if that land use had not taken place it is impossible to tell and pointless to debate, but its effect on the bastions of the ranges like Mount Torlesse and Mount Hutt was little short of devastating. The whole of the narrow road leading from Porters Pass to Lake Lyndon vanished as if it had never been and the only relics of its existence were one or two little islands of a few square metres rising on pedestals out of the raging torrent which flowed down to the bottom of Starvation Gully.

All the approaches to the bridges were washed out at one

end or the other and one, the Porter Bridge, collapsed under an avalanche of shingle which poured down on it from a great gully on Mount Torlesse. It was bad enough to lose the road but disaster struck the railway also, for a large part of the rail bridge over the Kowai River at Springfield disappeared and after the damage at the Waimakariri bridge the year before this was another major disruption for the West Coast line.

We had fairly heavy rain at Grasmere—enough to make unbridged streams impassable but not to do major damage. However, we heard that Dick Turnbull, one of the owners of Mount White Station, had been on his way home to Timaru when the storm brought the Hawdon River down in an old channel that cut off access to the Mount White Bridge and that he was marooned on the wrong side of it. We went to the rescue and got him safely to Grasmere, only to hear that there would be neither road nor rail to take him home for several weeks. He was not anxious to stay as our guest for all that time nor did he want to go back to Mount White, even if we had lent him a horse to ride there; so we offered him another kind of mount. We had an ancient and primitive bicycle with a fixed wheel and strange high handlebars which the cowman used to use to ride down to the paddocks to fetch his cows which grazed in summer time down by the lake. It was rather a fearsome steed which needed an experienced rider, as they say of restive horses. Once you were on it was not so bad but it took an active man to mount it. We called it 'the cow with the crumpled horn' after the old nursery rhyme. With this machine on the back of the truck we drove poor Dick as far as the first washed-out bridge at the Porter River. Here we deposited him and, rolling his trousers up as far as they would go and taking his bony horse under his arm, he set off to wade across its brown and foaming waters. His legs were obviously not accustomed to the open air—they shone white and spindly against the dark torrent and we watched him anxiously as he tripped and stumbled among the boulders and heaved a sigh of relief when he emerged safely on the far shore. Then, with a cheerful wave, he climbed on a rock and mounted his unfamiliar conveyance and we saw him labour up the long hill towards the unknown. Next time we saw him he remarked rather sadly, after thanking us profusely for our help, 'I think I carried it a lot further than it carried me!'

The road was not expected to be open for a month, the railway would take much longer than that to restore, and in three weeks our children were to come home from boarding school. How should we get them home? There was only one way to do it and that was on the backs of horses, so I made sure five horses were well shod and was all prepared to go to Springfield with four empty saddles and ride back again with children all aboard. However, it did not have to come to that because the road foreman said that if we promised to keep it quiet so that nobody heard that the road was open he would make a track for us to get in and out with our station truck. The children were rather disappointed. It would have been fun, they thought, to ride through like the pioneers a hundred years before.

The marks of that flood still deface the hillsides where it struck them and have served as a warning to justify the retirement of land of that kind from grazing. One thing is certain, Lake Lyndon has never been quite the same since. Its level is very low more frequently than before and the lagoon at the north-west end is seldom full.

In 1952 the Waimakariri Gorge had a visit from one of its most notable children, a man whose achievements are better known in the wide world than in his own country. Brian Brake was brought up at Arthurs Pass as a child of the storekeeper, Jack Brake. He became a photographer of world renown and only occasionally found time to come home and enjoy the mountains among which he grew up. Brian was keen to get some authentic pictures of kea attacks and I promised to let him know if I saw a chance of filming such a scene. There was snow quite early in June that year and the keas were troublesome both on the Burnt Face and the Bealey Spur. We had call-birds out in cages on both these blocks and went up to visit them from time to time. We found a small mob of sheep trapped on a narrow spur on the Burnt Face and instead of tramping a track and driving them down at once we decided to leave them, put a call-bird near and tell Brian to come and bring his film camera.

His gear included a large battery and a tripod and camera of considerable size—too much for any one man to carry. It was not long after I had been operated on for cancer of the bowel, and although I had got back in a remarkably short time to walking on the hill, riding a horse and leading a normal life, I wasn't fit to carry loads up 1000 metres of steep mountain side. To help us out I got Richard Johnson from Mount Torlesse to act as packhorse and we all took skis because the snow was deep up there. The day was fine and the sun shone as the three of us set off up the steep bush track which begins behind the old Cora Lynn homestead. The activity of my young companions was curbed by the weight they had to carry and I was able to enjoy a leisurely climb through the sun-spattered bush. Snow still clung to the branches of the trees as we got higher up, bowing their upright branches down with the weight of its burden. Now and then a branch would spring up, dropping its load with a heavy thud on the ground beneath and flinging a thin veil of silver mist into the sunlit air.

After an hour's climb the track emerges into the open where the fire which turned this forest face into a grazing block for sheep ended its destructive work. Above this point a few tall sentinels still stood, blackened and silver-grey, and great roots and fallen giants still lay about the slope under the canopy of snow. Nearly 400 metres above us I could see the grey smudge on the white surface which marked where the little group of wethers huddled on a hard-stamped island. With our skis on, and seal-skins to help us climb, we set out on a long traverse which would bring us out on a level with the sheep, but as we moved slowly up a thick grey mist began to form round the summit of Mount Bruce, which rises in a delightful sharp peak above the grassy face below. The higher we climbed the further it crept down, swirling and eddying, breaking and reforming; and coming ever nearer to the target of our expedition until it finally enveloped it and hid it from our sight. We climbed on, sliding our skis diagonally up the face—left, right, left, right, left, right, until we too were swallowed up in the blinding vapour. We went on because it might blow away as suddenly as it came, but it only got thicker until my companions could hardly see the phantom figure which preceded them and only my long years of walking and skiing on the place could tell me where I should guide my steps.

When I thought we were high enough we stopped and held a council of war. The mist was as thick as ever and the cool south-west wind which was driving it was not likely to go away. We had climbed, I reckoned, slightly above and well beyond the sheep, so that if we skied back and only gently down we might come upon them. Photography was out of the question but at least we might see if they had been attacked and if the traps which surrounded the call-bird had caught any keas. We took off our skins and I led off through the fog blanket. It was a weird feeling; impossible to tell whether the ground in front sloped up or down; impossible to adjust one's balance to a sudden drop when the skis shot away from under one or to a sudden rise that threw one head-first into a hidden snow-grass mound. Sometimes we left the ground altogether, wondering whether we should touch it again 9 centimetres or 90 below. It was all very well for me, with no load to carry, but it was fortunate indeed that both Brian and Richard were first-class skiers, for keeping their balance with those heavy and clumsy loads must have been almost impossible.

In spite of my vaunted knowledge of the ground we missed the tiny island where the sheep were trapped and by the time we realised that we were below them it was too late and too dangerous to go back. The fog had blanketed the face right down to the bush edge where the track went in and it was not until we were half way to the bottom that we emerged again into the sun.

Brian was philosophic; a cameraman has many fruitless expeditions no doubt, and though I promised to let him know if the chance occurred again before he left, rain washed that lot of snow away and released the sheep and the chance did not come again.

Brian came back to visit us once more in 1960 when he and Maurice Shadbolt were working on *New Zealand: Gift of the Sea*. They stayed with us and Brian took many photos; they used some of the material they got in that very successful book. More important to us was a long evening with two most interesting men which we shall always remember. Maurice remembered it also, for there is a short story about the high country in one of his books, *Summer Fires and Winter Country* I think, which contains a suspiciously familiar picture of our household customs.

Chapter 14

A DREAM COME TRUE

> Knowledge is of two kinds. We know a subject ourselves, or we know where we can find information upon it.
>
> SAMUEL JOHNSON

EVER SINCE the first meeting of high-country men at Tekapo in 1939 we had nursed the dream of a research organisation which would devote itself to the unique and special problems of the mountain lands traditionally devoted to sheep farming. The need for such an organisation was reinforced by the passing of the Soil Conservation and Rivers Control Act, and the campaign which followed it demanding a change in our farming practices. The Department of Agriculture from time to time employed an enthusiastic extension worker with a special interest in high-country pastoral farming; men such as Tom Sewell, who had worked at Grasmere during his agricultural degree course at Lincoln College, and Harry Sievwright whose efforts inspired the Mackenzie Country runholders after the war, but they were lone hands whose dedication brought them little prospect of advancement in their department and something much more cohesive was required. We needed a body of men whose hearts were in the high country but with a physical establishment in some central place where they could have a corporate entity.

The obvious choice for the location of such an organisation was Lincoln College, whose position in agricultural education and research was reinforced by its endowment lands in the high country, including part of the Mesopotamia Station made famous by Samuel Butler. Fortunately for us Lincoln

A DREAM COME TRUE 157

College had for many years arranged visits to the high country as one of the regular features of its degree and diploma courses, so not only the professional staff but many of the best of the students developed a life-long interest in the high country and its pastoral life. Grasmere was one of the properties which had at least one and often two annual visits from bus-loads of interested students. They often asked interesting and constructive questions and their attitude towards soil conservation gave us many opportunities to refute some of the more flagrant distortions of the campaign against the wicked sheep farmers. However, they were not always so serious and I remember one visit when we were passing from the woolshed, where I had been talking, towards the buses for their departure, and one rather gullible student asked how we used 'that snow-rake'—to which he pointed. This was a set of seed harrows to draw behind a drill and his mates had obviously kidded him into asking the question in order to have a good laugh at his expense. 'Snow-raking' is a general term used for any operation to rescue sheep from snow but it does not involve the use of a rake under any circumstances.

After several years of discussion and negotiation a conference was convened at Lincoln College in May 1951 to which were invited farming, Government, local and other bodies interested in high-country research. W. H. Gillespie M.P., who was Chairman of the Lincoln Council, took the chair and after the usual period of weighty and serious debate the following resolution was passed: 'That a committee from this conference be set up to enquire into the scope of existing activities, avenues for expansion, and to bring forward further proposals for the development of high-country research.'

The interesting thing about the committee which was appointed is how closely it paralleled the eventual composition of the management committee of the institute which was ultimately formed: three representatives of the High Country Committee of Federated Farmers, and one representative each from the Lands Department, the Department of Agriculture, the Soil Conservation Council, the Department of Scientific and Industrial Research, the Meat and Wool Boards'

Economic Service and Lincoln College. It was a good coverage of interested bodies and we faced the first meeting of the committee with high hopes. Unfortunately these hopes were rather rudely dashed when the meeting took place on 6 November 1951. After having elected me as chairman, the representatives of the Government departments announced with one voice that although they had been permitted to attend and join in the discussions they were on no account to agree to anything without reference to the heads of their departments. That left us a bit hamstrung from the start. They were all good, intelligent, willing participants but I knew only too well the suspicious inter-departmental jealousies which could put a veto on anything we decided to recommend.

The second meeting rammed the point home even more firmly. The Department of Agriculture indicated that, as the resolutions of the committee involved Government policy, as public servants its representatives were not in a position to express opinions unless asked for through the appropriate channel. The Lands and Survey Department followed with a similar statement and so the stage was set for a series of inconclusive discussions. I found myself in the position of chairman of a committee many of whose members were not at liberty to express an opinion! There was only one thing to do and that was to address ourselves to their political masters and stalk the corridors of power in Wellington. It was resolved that the invaders of the sacred precincts should be Dick Bevin and myself, and our appointment was with the Caucus Committee on Agricultural Research. I couldn't have been more fortunate in my companion because he had been for some years at Lincoln College and became the first director of the Meat and Wool Boards' Economic Service. He was also an old friend and a very charming companion so I looked forward to this expedition without too much apprehension, though I well knew how much hung on the impression we made.

I do not now remember much about our interview or even who the members of the caucus committee were and it is of no consequence anyway. What I do remember very vividly was our return journey from Wellington to Christchurch. It was in the days when the internal air services used mainly Dakotas and long before Wellington airport was more than a take-off place for light planes—if the wind was right! All flights were to and from what the New Zealand Broadcasting Service calls 'Para-para-oomoo' and everybody else, including Maoris, calls 'Paraparam'. Our flight was the last of the day and it was already a dark midwinter night. Droning peacefully southwards discussing the success or failure of our efforts during the day, Dick and I were rudely interrupted by an announcement from the pilot that there was fog at Christchurch airport and we were returning to Paraparaumu. The plane banked sharply round and set off in the opposite direction while the passengers speculated on whether the airline had not been well aware what Harewood was like before we left.

It was our fate to be accommodated in a small and rather primitive hotel at 'Pram Beach', with the threat of a pre-dawn departure hanging over our heads. Sure enough, we were called in the chilly dark, and mustered, unshaven and bleary-eyed, for a cup of tea about 6 a.m. Ugh! Off we set again on our southward course and this time they let us go the whole way. What did we see when we got there? An impenetrable sea of cotton wool tinged on the top by the first flamingo pink of the rising sun. The pilot again pronounced our fate, which proved to be Ashburton, which boasted a minature landing ground where a DC3 could get in.

'Hoorah for the displaced persons' flight!' cried Dick. 'Why don't we ask him to take us to Mount Cook? It'll be a lovely morning there and only a little further.' However, the pilot soon vetoed any little jaunts like that and down we popped, skimming one gorse hedge with our wheels and stopping just in time before another loomed up.

Breakfast at the new Devon Hotel. A proper wash, good food and hot coffee; that was better, and the rather silent and gloomy passengers became quite gay and talkative. There were only about a dozen of us and after breakfast we were offered a choice. Stay with the plane and hope the fog would clear, or be driven to Christchurch in a taxi at N.A.C.'s expense. Some wag suggested that the pilot put his Dakota on the Main South Road and drive us up in comfort but the suggestion wasn't received with much enthusiasm. Even a Dakota would spread its wings across the whole road! Most of us stuck by the ship and we were rewarded about 1.30 p.m. by the sight of power pylons at Yaldhurst, a few feet, it seemed, beneath our seats; then a slowly spreading rent in the cotton wool let us sneak down to land. Our poor wives wondered where we had been!

It was February of 1953 before we knew the result of our efforts. The Minister of Agriculture advised that Cabinet did not propose to do anything in the matter of setting up a high-country research station. Now we really were displaced persons with nowhere to go! Our third committee meeting had to decide whether to give it up or keep on trying. To its eternal credit it resolved to battle on against all the obstacles which beset its path. At this stage I resigned from the chair—not in disapproval or despair—far from it, but because Mary and I saw a long-hoped-for chance of going to England. Leo Chapman, who had been a leader of the enterprise from the beginning, took my place as chairman.

It was two years before we had any reward for our efforts to get high-country research under way and it came from an unexpected quartet. Our work inspired the Soil Conservation Council to contemplate doing themselves what we had advocated from the start and they launched their proposal at a conference at Kurow in January 1955. It was to be a partnership with the South Island catchment boards and there was no proposal for any participation either by runholders or other Government departments, in fact the runholders' request for representation was firmly rejected.

At this point we received a gift from heaven in the shape of a man who was clearly of divine origin. Dr Merton Love, a specialist in range management at Berkely University in California, came to work for a time at Lincoln College. He was not only an expert in his field in America but he was a warm and friendly human being who immediately understood the runholders and their point of view, and saw how and where research could best be conducted. After six months at

Lincoln College he wrote an outspoken and authoritative report recommending an institute based at Lincoln with a research station located somewhere in the high country. Armed with this powerful weapon the High Country Committee and Lincoln College combined to attack the authorities again and succeeded in getting a committee of the Soil Conservation Council—there seemed to be committees galore in all this work—and representatives of the Lincoln College Board of Governors to meet at Lincoln and discuss Merton Love's proposal. At this meeting, which was presided over by Charles Hilgendorf, a fateful resolution was passed: 'That to bring about co-ordination of work in the overall field of investigation concerned with this problem, a Tussock Grasslands and Mountain Lands Institute be established.' It also resolved that its work should be under the guidance of a management committee set up by the Soil Conservation Council and that it should be located at Lincoln College. After all the years of patient but determined negotiation we had got our research institute.

However, things do not move very fast in public affairs and it was another three years before the institute was finally launched and even then the jealous departments insisted on the deletion of the word 'research' from its title for fear of it competing with their inalienable right to such work.

However, the long-awaited day came in 1960 with the appointment of a management committee presided over by Mr R. M. D. Johnson who, as a member of the Soil Conservation Council and Chairman of the North Canterbury Catchment Board, had worked tirelessly to get the scheme approved. The first director, L. W. McCaskill, was appointed and the institute was under way.

I do not propose to trace its history here, but its very existence and the dedication of its staff have had an enormous influence on the attitude of high-country runholders and on public knowledge and understanding of their ancient profession.

When I look back at the days when changes were being forced upon us I am amazed at how our outlook has altered.

The work of the institute has been a very important factor in this change because, apart from its other achievements, it has collected and disseminated the knowledge of which we now make so much use. I contrast our early meeting at Lake Tekapo in 1939 with two significant ones in 1969. One was the high-country field day held at Lake Hawea in January 1969, which ended with a highly technical discussion on trace-elements, in which the local runholders took a full part. Thirty years before they would hardly have understood a word of it. The second was a meeting of our High Country Committee at Broken River in February of the same year. Here the discussion ranged over improvement work undertaken with and without Soil Conservation Council subsidies, and the impossibility of retiring, by physical divisions, all areas of Class VIII land. Even the conservatives among us were prepared to accept that changes were possible and might even be profitable. A few years before, the catchment board officers present would have been verbally torn apart limb from limb!

Chapter 15

BOMBO

> Your sheep, that were wont to be so meek and tame, and so small eaters, now, as I hear say, be become so great devourers, and so wild, that they eat up and swallow down the very men themselves.
>
> SIR THOMAS MORE, *Utopia*

WHEN OUR CHILDREN were young, like most other country families they had pets: rabbits—perish the thought—guinea pigs, Canada geese—Robin had a much-loved goose called Angus—and of course pet lambs. The greatest of these was Bombo. Bombo became a monster, but worse than that, or rather because of that, he became quite dangerous. An unsuspecting human with his back turned was a fatal attraction; Bombo would stand back 10 metres to get up speed and launch an attack like a prehistoric battering ram with 330 kilograms of solid mutton preceded by a skull with the texture of a granite boulder. The boys got good at playing him like a matador in Spain but this probably made him worse; anyway the average adult was not so nimble, and we were afraid somebody would get badly hurt. We finally banished him to a large paddock near the railway yard at Cass inhabited only by the bulls, a few large horses and occasionally the rams, all of which we thought could quite well look after themselves. Unfortunately this sanctuary was sometimes invaded by women from the railway houses looking for mushrooms, notably by our postmistress Aunty Grace Robertshaw and her bosom crony Ada Clausen. They were a delightful pair at any time and formed a sort of comic team rather like two female Ronnies or the Gert and Daisy of long ago, scoring hits at each other's expense and commenting on the vagaries of their neighbours and their husbands and the guards and drivers who got the benefit of their barbed sallies when the trains stopped at Cass. Ada had the advantage of a Midlands English accent and when she came to Grasmere to pick raspberries she always used to say, 'I've brought me jum pun' (jam pan).

Well Grace and Ada went mushrooming in the Swamp Paddock—with the jum pun no doubt—and the consequences were disastrous. Bombo saw them from afar; two well upholstered ladies, heads down and unsuspecting, and the temptation was too much for him. He scored a right and left before they knew what was happening and sent them flying for the shelter of the railway fence through which they squeezed with great difficulty while Bombo butted and stamped like a wild bull of the pampas. We were not popular after this as you can well imagine, and rather than offend our lifeline—for Aunty Grace was our only link with the outside world—we banished Bombo even further; to Cora Lynn, in fact. Here he resided on the banks of Broad Stream which in those far-off days was unbridged and often turbulent. Bombo had a fellow resident there, for the Public Works Department had built a little cottage on the north-west bank so that a man could always be available if motorists got into difficulties with the flooded stream. Stan Ashby occupied it for many years, until the present bridge was built. Stan and Bombo soon established a state of armed truce; Stan never went to the creek without his shovel and Bombo learnt a healthy respect for this iron-shod weapon which proved to be even harder than his head. Whenever Stan was on the scene all was well, but any unsuspecting motorist who got out of his car to look anxiously at the racing waters was fair game if he happened to be on the villain's side of the creek.

In 1952 the winner of the Melbourne Cup was a New Zealand horse called Dalray which belonged to a Greymouth businessman called Neville. After his triumphant return from Australia he and his wife were driving back to the West Coast and, as so often happened, Broad Stream was in a state of minor flood. Its swift, discoloured waters made it difficult to see where the ford lay and whether it was safe to attempt a crossing, so the couple got out to look. A fatal mistake; while they looked at the surging water trying to decide where to make the crossing, completely disregarding the harmless woolly creature standing on the shingle, evil thoughts were passing through that innocent-looking head. The lady's back view was too tempting a target and, without warning, he made a charge, sending her head-first into the water. Stan came on the scene after she had struggled out, and the injured pair demanded that he tell them who owned this vicious beast so that they might be prosecuted—but strange to say he had no idea!

When I first went to Grasmere vets were few and far between; there were no vet clubs because both sheep and cattle were of such low value that farmers did not think them worth spending money on. Outside Lincoln College and the Department of Agriculture the only independent vets were ones concerned with racehorses and pet dogs and it was not until after the war that a serious policy of training vets for farm animals was adopted. The trainees had to go to Australia to study, and when they came back fully fledged veterinary surgeons with degrees they went out to staff the newly formed vet clubs, often with fairly limited practical experience.

Most back-country men are used to the rougher type of surgical work on animals and usually did the castration of unbroken horses, as they did their lambs and calves, with reasonably good results. However, when there was a trained vet available at Darfield and we had a colt to do I thought it worth getting the newly appointed man to perform the operation. We had a great big gangling chestnut colt which had not even been roped and tied up so he would need considerable care in handling if he was to be thrown and castrated. When the vet arrived all the men were on hand to help. We had got a rope and halter on him by this time, and with three or four men to hold him we led him out on to a dry grassy spot in front of the house. Very professionally the young man got out his bag of instruments and arranged them ready for action. They included a very new and obviously unused anaesthetic mask of bright green canvas. It had been difficult enough to get a halter on the brute, let alone this canvas nosebag, the very sight of which sent him into a paroxysm of fear; however, in the end it was done and a liberal dose of chloroform from a bottle was poured on to the sponge which it contained.

'Right,' said the young man confidently, 'just lead him round a little and he'll soon go down.'

So the stalwart shepherds led him, or dragged him in a circle. He was an ungainly creature with big feet and a rather curious high-stepping walk, and we all watched as he went round once, twice and a third time.

'Strange,' said the vet, 'that much dope should have put him out by now.'

He got out the bottle again and before the fourth circle he poured another good measure into the mask. The bottle was more than half empty by now and he eyed it rather anxiously, I thought. The circle began again and this time indeed we could see the horse's gait was more uncertain; he did not exactly stagger but his massive hoofs made unpredictable manoeuvres in the air before he put them down. Round he went again but there was still no sign of the expected collapse. The poor vet was getting really anxious now. Not only were there five or six critical witnesses whose ill-concealed grins must have been very galling but up on the roof of the house sat our daughter Anne entranced with the performance; worst of all, perhaps, the vet's young wife had come with him for a ride into the unknown high country and was watching every movement in the drama from their car. There was only one thing to do and he did it; he emptied the last drop in the bottle into the mask and I'll bet he breathed a silent prayer as he did so. The animal got clumsier, I admit, but in spite of his uncontrollable feet he continued to plough round and round behind the men who dragged him. Drugged he certainly was, but down he would not go. At last our head shepherd, Don McLean, lost patience with it all. He pulled the rope round behind the animal and dropping it into the hollow between the fetlock and the foot he gave a mighty heave and, before the horse could turn, his hind legs doubled under him and he went down with a crash. After that there were no further problems; he was soon hog-tied and unable to resist even if he had not been in a state of stupor. The poor vet was so eager to complete his job with professional skill that it was soon done. The amount of chloroform the horse had inhaled resulted in a slow return to normal and it was half an hour before we got him on his feet. I have no doubt that he was an unusual horse, and I felt very sorry for the poor chap who had the bad luck to strike one like that for his first attempt. It shows how important experience can be.

Lack of experience or just plain carelessness nearly cost me dearly in 1956—that awful year of drought and fire. It was near the end of shearing in January, and the days were searing hot and men and dogs were tired out. I had told a young shepherd to take a large mob of sheep to the Ewe Country; a drive of about 8 kilometres, half of which had to be through a block of rather scrubby country where the sheep could easily straggle off the track if they were not held fairly firmly in a bunch. As there were 700 or 800 of them this was a test for tired dogs. At the last moment I relented a little. I had a bit of time to spare and I told the lad I would bring a couple of dogs in the car and help him over the difficult part, because I didn't want him to lose any of his sheep on the way.

The drive was slow and, as I expected, I had to use my two dogs a good deal to hold the lead of the sheep. My good heading dog, Kay, was in his prime and would never let sheep go once he had headed them, and the other was a big black huntaway called Boss whose rather ruthless style made him ideal for handling a large mob. Both of them had had enough by the time we got the sheep to the Ewe Country and I was more than thankful that I had gone, because I didn't think the boy's dogs could have held them without my help.

When I got back to the car I opened the boot, the panting dogs jumped thankfully inside and I drove home. The shearing shed claimed my attention as soon as I got back, and putting the car in its shed I went straight there, forgetting all about the two dogs in the boot.

It was several hours later that I had reason to go to the garage for something and the first thing I heard was a strange snoring noise. Only then did I remember what I had done, and flinging open the boot I found two completely prostrate dogs. Kay had forced his nostrils into the crack where the boot was divided from the back seat of the car and was clearly getting a little air through that tiny aperture but Boss was lying flat on the floor and at first I thought that he was dead. I lifted the two of them out and fetched a hose and doused them with cold water. Kay recovered fairly soon and I was thankful to see him struggle to his feet and stand swaying groggily and gasping in great lungfuls of air. Boss's heart was still beating but there was little sign of breath and I pumped his ribs to try to make him breathe. At last a sort of gasp came out of his open mouth and very slowly the lungs began to work again, but it was several hours before I was sure he was going to live and several more before he too staggered to his feet.

Of course they had been exhausted when I put them in and the heat and lack of oxygen soon caused them to succumb. If I had not had to go there I might have forgotten them until their feed-time at night. Boss would certainly have died by then and very likely Kay also. Both dogs served me another ten years, which I did not deserve.

I have always said that fishermen never caused us any worry; they confine their activities to the lakes and in most cases are considerate and well behaved. There is one way, however, in which they can be thoughtless and careless; they eat tinned food and leave the tins about. That doesn't sound a very dreadful thing to do; bottles sound much more dangerous, but cattle are curious creatures and this is where the trouble comes.

Some years ago I was riding round one of the blocks when I saw a young steer so thin that you could have 'cut your fingers on his spine'. When I got home I told the head shepherd to go and shoot it for the dogs, though there was little enough meat on it for dogs to eat. 'Bring me back the head,' I said, 'it looks to me as if it might have wooden tongue.' Wooden tongue is a disease which cattle can pick up through pricks or mouth wounds; it can be spotted by swellings on the side of the face above the mouth. They brought me the head to look at, and there were the characteristic swellings; to my surprise they were caused not by wooden tongue but by a flattened tin jammed between the animal's back teeth and projecting into its cheeks. The gullet was almost blocked by this obstruction though the presence of unchewed grass in its mouth showed that it had still endeavoured to eat with its front teeth.

That was the first of several. In two years I think we had five cows with different sorts of tins stuck in their mouths round Lake Grasmere. One heifer had a hole cut through her cheek by the sharp corner of a tin and others were only saved by somebody noticing it in time. It became necessary to do something about it and we resolved to donate a couple of acres of land on the shore where most of the fishermen camped, for a reserve which could be fenced in and used exclusively by them. For many years discussion had gone on about access to the lakes. Strange to say when the university endowments were made in 1873 the draughtsmen forgot to reserve the lakes and their chain surround to the Crown and so the university became the owner of the lakes and their shorelines, which they leased to their tenants. Our freehold land also extended to the water's edge, so officially there was no access to the lakes themselves for fishermen.

It took some years to rectify the mistake; for the Public Service, like the mills of God, grinds slowly; surveys have to be made and Ministers' consent obtained and the rights of individuals like ourselves protected, so the final transfer of the 'lake chain' to the Crown more or less coincided with our donation of the fishermen's reserve. We also donated the beautiful strip of mountain beech bush on the far side of the lake as a scenic reserve, to protect it from our own increasing cattle which used to wander through it. The Lands Department paid to have it fenced and also the fishermen's camping ground, and in exchange for the freehold we surrendered round the lake, they gave us an area of freehold land at Cora Lynn to fill in a gridironed area behind the old homestead.

Considering the large number of birds throughout the Southern Alps it is a strange thing that comparatively few keas' nests are found. They were always supposed to nest in very wild and inaccessible places, rocky caverns which were hard to approach, and there have been authentic records of such sites which have led people to believe that they were used exclusively. Peter Newton exploded that theory many years ago when he was hunting keas as a winter job and found nests in easy, undulating country, as he records in *Wayleggo*. They also nest in the bush, it seems, because Lance McCaskill and a party found a nest near the bush track that leads to Lake Minchin up the Poulter river. Newton claims, as others before, that the birds nest in the winter and this I believe is usually true because young birds are mostly seen on the wing in spring and early summer but the Minchin nest was found in January so it is clear they may nest twice a year or over a long period.

The best authority on keas that I have met is Dick Jackson, a deeply dedicated observer who frequented Arthurs Pass and the Waimakariri Valley in the fifties and early sixties. Dick, like many other bird lovers, just could not bring himself to believe that his feathered friends were evil murderers. So anxious was he to prove their innocence that he offered to pay me the value of the extra sheep we lost if I gave up killing keas for a year. It was a noble and very reckless gesture because a flood or snowstorm or a piece of careless mustering could cost us many hundred sheep which nobody could prove had not been killed by keas, so for his own sake I refused to accept the test. By sleeping out at night and watching and using the natural inquisitive nature of the birds Dick became familiar with a number of breeding pairs, finding their nests and determining the range of their territory. His dedication can be measured by a remark he made to me once.

'I never sleep in huts,' he said. 'If you're in a comfortable warm bunk in a hut and you hear a bird outside you won't get up to look at it. If you are lying on the hard ground beneath a tree, uncomfortable and cold, you will jump up gladly and observe its behaviour.'

Dick led me to the only keas' nest that I have seen and taught me a useful method of attracting them at the same time. He had been up at our Top Hut on the spur of Powers Country where keas are a common sight, and having found a nest he offered to show it to me. An excuse to go to the Top Hut is always welcome, so I was delighted to accept.

The date was 31 October, and as we walked up the spur above the hut there were still patches of soft snow lying about. The nest was somewhere in the stunted mountain beech bush where it dies out in gnarled and lichen-covered trunks whose half-exposed roots cling tenaciously to the rocky promontories of the ridge. It was well on in the afternoon by the time we got there and too late to take photographs and weigh the young birds, which had been the main objects of our expedition. Dick wanted me to see the behaviour of a nesting pair so we sat down to watch. The female was there all right but Dick said the male would not return—from the pub, as he put it—until just before dark. The hen is then supposed to flutter anxiously about her spouse and go through all sorts of affectionate posturings until he feeds her from his beak. Unfortunately Dad must have got drunk on this occasion and did not come home before dark so we had to leave the poor hen fluttering disconsolately about the rocks and go down to the hut and feed ourselves.

Next morning Dick was up at 5.30—probably cold, for he had little bedding—and we breakfasted hastily and left to find the nest. Fog began to drift up and there was a cold south-west wind along the ridge. I had brought two interested shepherds with me, Kevin Ryan and Karl Bollinger, and we three sidled round above the bush beneath the fog carrying a bag of cameras. Dick went up and on to the ridge to find an ice-axe he had left the night before and we foregathered in the fog where the spur bends south. At first we couldn't find the stone cairns with which he had marked the position of the nest, but in the end, scrambling among the stunted trees, we stumbled on a pile of stones. It was a steep and awkward place. Loose scree was sliding into the twisted tree trunks and was only held here and there by clumps of mountain totara and a few ragged coprosma bushes. The nest was just a hole under a big rock which was lodged against an aged half-dead tree and there were two compartments. In the occupied one, two huddled, grey, furry shapes blinked in the light of torches while the mother dodged about behind them in the gloom. Dick dragged them out unceremoniously with his ice-axe—now we could see what this was for—and we tried to photograph them in the gloom of the bush. He hung them upside-down to weigh them and put rings on their infant limbs. One was much bigger than the other and a third, which Dick had seen before, had disappeared, together with an unhatched egg.

We left them in their unsavoury-smelling hole and climbed up the boulder-rolling slopes to sit on the ridge. Up swept a young kea with a banded leg so Dick proceeded to stalk it and try to read the band. To tempt it close enough he produced a large lump of butter which he dabbed on the rocks to lure the bird towards him. It came, right enough, but so did an older male who promptly expelled the youngster and hogged all the butter himself. It would be easy to substitute the butter for the traps we used and add a little strychnine in order to get all the keas you want—so thought the shepherd members of our party, but they were careful not to speak their thoughts aloud.

There was no more to see, so we hurried down to the hut, loaded our packhorses and set off down the Bealey Spur to Cora Lynn with a spatter of south-west rain rattling on our oilskinned backs. Forty years in the high country and I had seen my first keas' nest. I was ashamed to have been so unobservant!

One day some years later Dick Jackson's butter trick was useful to me. When I woke up one winter morning and went out to the kitchen—Mary was away attending at the birth of another grandchild—I heard a scuffling on the lean-to roof of the store-room and went out into the snow which lay round the house. There was a cheeky-looking kea pecking at the icicles hanging from the roof. Here's a chance, I thought, to try the butter trick and see if I can get him within reach of a wire hook. I went indoors and got some bright yellow butter, which I cut bits off to tempt him. He got the idea at once and I lured him piece by piece as far as the kitchen porch. No, it couldn't be possible! First I threw the pieces in the porch and then on the doorstep itself, and still he followed eagerly. Damn it, I thought, it's worth a try! And I threw a piece right inside. At first he crept in with infinite caution and grabbed the bait and fled. I threw a piece further in and hid behind the door and when he was clear of it I slammed it to. He was mine now, though he took some catching in the kitchen and I was afraid he might break a window in his efforts to escape. He was probably a stray visitor from the ski huts along the range where scraps of food are easy to come by and hut doorways hold little menace. There are hundreds of these birds—mostly young—whose winter food supply suddenly multiplies a thousandfold in the skiing season. As long as they confine their attention to raiding refuse dumps and chewing skiers' gloves and hats carelessly left outside no harm is done.

One of my most delightful memories is of watching a kea on the back lawn at Grasmere skilfully removing from the line a pair of delicate scarlet panties belonging to my daughter-in-law, and carrying them to the top of the woolshed with such obvious glee that you could almost hear the chuckles. Nonetheless there are still the same disturbing numbers of rogue birds which turn to sheep for prey, like man-eating tigers, and have to be destroyed—if you can catch them!

Chapter 16

DISASTER—OR BLESSING?

A little fire is quickly trodden out,
Which, being suffered, rivers cannot quench.
                    SHAKESPEARE, *King Henry VI*

NEXT TO SNOW the greatest danger we faced at Grasmere was fire. With the railway bisecting the property from Craigieburn to Arthurs Pass there was a constant fear of accidental fires. Usually the danger was greatest in the spring in spite of the fact that this is often a wet period. The reason was that it is always a season of nor'-west winds, sometimes blowing for weeks on end, and the seed heads of last summer's grasses have been beaten down by snow into a dry and highly inflammable carpet as yet unpenetrated by new green growth. Steam engines are noble and impressive monsters and one can well understand the nostalgia which people feel for the wonderful era of the steam locomotive, thundering its way through the country with great gouts of smoke and steam epitomising its latent power. However, its exhalations often included a great deal more than smoke; hot ashes dropped on the track from punctured fire boxes and sparks shot out of its funnel to fly before the wind and land on some tinder-dry patch of last year's grass. The locomotives were supposed to have spark arresters to stop this kind of accident but these caused some restriction on the power of the engine so the drivers were never keen to report it when they were burnt through.

If a fire occurred as a result of such a fault the landowner could claim damages against the railway, but—and this was a major but—he had to be able to prove not only his own losses resulting from it but also that the engine was faulty. Needless

DISASTER—OR BLESSING?

to say this was impossible to do in most cases. The only time I knew the Railways Department to pay up and look pleasant was when such a fire occurred on Mount Bernard, next door to us. The owner in this case was already a Member of Parliament and was to go on to become Minister of Railways! He asked me to act as arbiter on his claim for burnt sheep—of which there were quite a number—and I took a malicious delight in including the loss of grazing at a fairly generous figure. As far as I knew the Railways Department had never before accepted liability for loss of grazing but in the spring it is a very real loss and I was only too happy to set a precedent for a claim which I might one day have to make myself. The fact that my neighbour's claim was accepted and paid for shows how differently influential people can be treated.

In 1956 we had a very dry spring and summer and by Christmas time the paddocks were parched and burnt and the mountain slopes were dry as tinder. Shearing was over early and we had sheep out on Powers Country before Christmas to try to spell the desperately dry winter country. On the morning of 21 January, after it hadn't rained to speak of for a month, we suddenly saw smoke blowing across the homestead from behind the cookshop. All round the western and southern side of the buildings the native vegetation had been overrun by one of New Zealand's worst weeds, broom, and it was from this area that the smoke was coming. Immediately there was a wild rush—to see what danger threatened and to take steps to protect the homestead buildings. One man with great presence of mind saved all the dogs; for in that belt of sheltering scrub four teams of dogs were chained securely to their kennels. Racing from point to point through the blinding smoke Eldon Coates, who was one of our musterers, released about thirty dogs, whose loss would have been a major disaster, for without them we cannot work a station. With a fresh nor'-west wind increased by the suction of the rising air above the flames the fire tore through the dry scrub like lightning, and by the time the last chain was loosed the flames were close upon the rescuer. Only two pups in a run by themselves were lost. The musterers' hut and the shearers' quarters were directly in the line of the flames; the dry grass between the buildings and the scrub and pine trees at the back were no sort of firebreak.

I need not describe the period which followed, nor I think could anyone remember what happened while we were all trying to save the buildings and protect the house. There was no time for orders and planning; everyone did what he saw before him to do, with hose or bucket or makeshift beater and it was not until the flames had swept past, leaving scorched paint and smouldering woodwork that we realised the equally great danger to the hills behind. Once on an open tussock hillside the flames roar up a face as the rising air drives them on, and 400 hectares of winter country lay at the mercy of this evil demon. We were too late to stop it getting on to the hill and I doubt whether we could have done so if we had let the buildings go. Now there was nothing but Ribbonwood Creek between the fire and Lake Pearson 8 kilometres away. That day remains a nightmare in my mind; blank periods are interspersed with vivid memories, some horrific and some incredibly comic. At one stage I had organised a gang of men, mostly railway employees—for people turned out from far and near—and left them working on a ridge where the grass was thin to stop the fire back-burning towards the Cass River Valley. The task was not difficult because the wind was behind them and all they had to do was keep pace with the fire as it climbed towards the shingle, where it would die out. No sooner was my back turned than one man decided it was too much like work and they needed a rest. The rest they took cost a gang of men a gallant fight which lasted half the night in a scrubby gully further back towards the Cass.

Racing down the hill to the homestead I was greeted with the news that the fire had swept right across the Bailey block and crossed the fragile barrier of the Ribbonwood Creek. I had a brief glimpse then of the unforgettable picture of our good neighbour, Gerald Urquhart, a man of huge stature and truly rugged structure, dressed in a half-burnt shirt and with his almost bald head covered by my wife's huge straw gardening hat tied on with a scarf. Leaping into a car I gathered a few men and took them to Lake Pearson to see if we could find a place to burn a break and stop the fire before it got to Flock Hill. At Lake Pearson another strange sight greeted me. We had arranged long before that the annual picnic of the Clan MacLeod Society should be held at Grasmere on that day and in the stress of the fire all thought of this arrangement had been driven from my head. The bus, with its cargo of young and old MacLeods, had arrived at Lake Pearson, where they were to bathe before coming on to lunch at Grasmere. Smoke in great billowing clouds greeted their astonished eyes and when smoke was followed by a menacing wall of crimson sweeping over the fan towards the lake they hastily retired into the water where they stood like storks, knee-deep in its chill protection.

I organised a fire break there—which failed in fact because it got away from the gang, which was too small to control it—and then I hurried back to the distant homestead to see what had happened there. We had a crop of oats which some men were trying to save and they had got one of our tractors and a plough to try to plough a furrow between it and the fire. Alas, the fire came too fast and the driver had to leave his seat and run, leaving the tractor to its fate. Its fate was three burnt tyres and the loss of all its paint; fortunately it was a diesel tractor and the fuel did not explode. Here I saw another strange vision: a woman whom I had never seen before, her dress tucked bulkily into a pair of pink bloomers, beating at the advancing flames with all her might. When I asked her where she came from she said, 'Oh! I was driving along the road and saw the fire so I came to help.' Bless her heart! I still don't know who she was.

All the Mount White shearers dropped their tools and came over 30 kilometres of winding road to help, and many others from as far away as Springfield were there as soon as the news got round.

Nothing could save the great tussock slopes of Mount Bailey and the fire consumed every living thing thereon, except for one spur which was cut off from the rest by shingle slides. Fortunately there were no sheep on the block, or at least only a handful, but the tussocks were so dry before the fire that they burned to their very roots, leaving a cup-shaped hollow in the ground where they had been. As the fire died out the weary gangs collected at the house where Mary had stayed throughout, minding the telephone and preparing food and drink even when the flames were only a few chains from the kitchen door. Her only attention to our own interests was to collect her jewellery from our bedroom and put it in the pocket of her skirt. The organisation of fire-fighting is a most difficult exercise and a central point of command and communication is essential; one devoted hero or heroine must stay and man it no matter what happens. Active control is very difficult always and in those days we had no walkie-talkie sets to keep contact with distant fronts and no time to organise gangs with someone in authority to command them. Fortunately station managers and head shepherds are accustomed to controlling small gangs of men and when they are available there could be no better leaders.

Somebody sent a truck to the Bealey Hotel for beer and as the drama and the trauma died everyone stood around with mugs of beer recalling incidents of the day. Among them were two which concerned our old cowman, Percy. Percy's hut, which no one cared to share because cleanliness was not one of his virtues, was one of the first threatened by the fire and when his few belongings and his bedding had been removed—some said they would have been better left to burn—he obviously felt that all this excitement was beyond his power to cope with, so he retired to the garden and at the height of the drama Mary found him on his knees with a pair of hand shears clipping the edge of the lawn. Talk about fiddling while Rome burned!

There was worse to come though, as far as Percy was concerned; when all the dogs were loosed they headed for the killing house where scraps of meat were often to be found, and near by was Percy's hen run. Some of the younger dogs, finding not much to eat, solaced themselves by chasing and slaughtering half a dozen of Percy's cherished chooks and scattering the rest all over the adjacent hill. Even that was not the sum of his misfortunes though, for when it came to milking time and he rounded up his four or five cows, every bucket on the place had vanished completely. Percy's rage was horrible to hear—there was no possible justification in his eyes, fire or no fire, for taking his buckets from the cowshed!

There was little left to do by five o'clock, except that up in the gully behind Little Bailey the fire had got back from the ridge into a patch of high scrub in which nobody could approach it. Nowadays this sort of place could be dealt with by a 'monsoon bucket' from the air but in those days there was nothing to do but sit round it far into the night and prevent it from spreading.

Our examination of the hill next day showed the devastating effect on the vegetation. The tussocks appeared to have been completely destroyed; and would the other grasses ever grow again? Here and there smoke rose from places where plant roots were still burning underground and in one or two wet places patches of peat and moss were smouldering steadily. The rest of the hill was still and black and looked utterly and completely dead.

Mary and I had arranged to go to Christchurch the following day and as we drove up the hill towards Ribbonwood Creek I looked across at the blackened hill. Halfway across it a plume of smoke still rose from where some springs came out to form a little stream. I slowed and stopped the car.

'Mary,' I said, 'I am not going off the place while there's still fire burning on the hill,' and I turned the car and drove home. Some power must have spoken to me in that moment for by five o'clock that afternoon the countryside was alight from the Cass Bridge to Lake Sarah along the railway line. A train had started five fires between those points and although most of them were put out by the railway staff, one on the Long Hill swept away completely out of control. Here we were with a nor'-west wind, tinder-dry tussock and scrub and 30 kilometres of open country ahead of the flames. This fire started late and we did not have hours of daylight to deal with it, so it became a most difficult night operation. To make matters worse one of the fires lit by this engine was at the point where our power line crossed the railway in its underground cable. The fire melted the lead cover of the cable where it emerged from the ground and burnt the insulation on the wires running up a pole. Power was cut off from the whole homestead and the work of preparing food and drinks for 100 fire-fighters had to be done by candle-light.

We sent a frantic appeal to Ted Salvesen to send someone up to mend the wires, and like the good friend he always was he sent a man racing to our help. This rescue served another turn also because our younger daughter Anne got a lift home with him. She was in Gisborne when she heard the news of the first fires and, unable to bear the thought of being useless and far away while we struggled in our agony, she caught a plane and arrived in Christchurch just in time to come up in Wooff & Salvesen's van.

Working in the dark and cold Ellis Dawkins restored the damaged cable and about midnight the weary women, still hard at work making sandwiches in the house, were rewarded by a sudden blaze of light.

By about 1 a.m. the battle appeared to be won, and for the rest of the night watchers in relays patrolled the hills, of which about 2 kilometres had been swept.

The next day the wind was stronger, and at about 1 p.m. roots smouldering beneath the surface, as they had on Bailey three days before, suddenly burst into flame and before anyone could get to the spot the whole conflagration was away again. The fire swept most of the rest of the long low hill and was only checked at the old cart track which crosses it from the north end of Lake Pearson. At one stage it was burning downhill towards the strip of mountain beech trees which fringe the steep shore of Lake Grasmere and I went to see what could be done to save this beautiful background to the lake. Here I found two men battling alone to subdue flames which were licking up scrub bushes 2 metres high, with no better weapons than beaters of wet sack. Their task seemed impossible and I marvelled that they would even try. Only men bred and dedicated to the country would have carried on against such odds. One was my own head shepherd Ivan Williams and the other Fenton Westenra, who now owns

Craigieburn in partnership with his wife. It has never ceased to amaze me that while I went to get other men to help they won the fight alone and when we got back they had saved the bush.

The battle went on till 2 o'clock the next morning, by which time the remnants of the fire were slowly burning down the steep face opposite Lake Pearson in scrub so high and dense that no attempt could be made to put them out. The wind had died and begun to change; cool puffs blew in our faces and the stars above were dimmed and blotted out; then the first drops of heavenly rain began to fall, then more and more and more, until our parched and scorched skins were drinking in the blessed moisture and the spitting crackling flames below died in the steady and persistent stream. We went home and slept—and slept—and slept; feeling at last that we were safe after five days of stress and strain and agony. And while we slept the blessed rain went on steadily and mercifully, washing the blackened ground and what was left of its once green covering of pasture so that when we awoke there was some hope that everything which once sustained sheep might not have been destroyed.

Chapter 17

REVOLUTION

> Behold, a sower went forth to sow; and when he sowed, some seeds fell by the way side, and the fowls came and devoured them up: some fell upon stony places, where they had not much earth: and forthwith they sprung up, because they had no deepness of earth: and when the sun was up, they were scorched; and because they had no root, they withered away. And some fell among thorns: and the thorns sprung up, and choked them: but others fell into good ground, and brought forth fruit . . .
> *Matthew*, 13, 3-8

IT'S AN ILL WIND indeed that brings no good to anyone and, although we thought our ravaged hills an unrelieved disaster, the cook who threw hot ashes in a scrubby shingle pit to start the Bailey fire and the driver whose engine started the second lot of fires turned a page in the Grasmere book and made their contribution to a revolution in high-country farming. After inspecting the damage on Bailey with me, R. D. Dick, Chief Soil Conservator of the Catchment Board, suggested that we apply for a subsidy to topdress and oversow by air 500 hectares of the burnt face of Bailey. As far as I know no area in the real high country anywhere approaching that size had been topdressed before. Lower hill country in both islands had been treated in this way with success but most of us still believed that the harsh climate, poor soils and short growing season in the mountains ruled out any chance of improving production significantly by such methods. I was extremely sceptical and doubted also if the money would be provided by the Soil Conservation Council for such a massive experiment, but in our situation anything seemed worth a trial.

To my surprise the board and the council both approved the plan and we found ourselves faced with what sounded like a very difficult project. There is no need to go into details; aerial topdressing is now so commonplace that it needs no description, but when you first see a pile of superphosphate lying like a white mountain in the corner of a paddock and one tiny little plane drawn up beside it you may be forgiven for wondering whether the notion of transferring that mountain to a real mountain side and spreading it evenly up to 1200 metres is not some kind of joke. The weather of course can foul up all your plans, and breakdowns and mishaps disrupt the work when the weather is good to you, but the incredible speed and efficiency of pilots and their staff seem to triumph over all the difficulties, and hour after hour at intervals of three or four minutes the little mosquito zooms in and turns, and in less than another minute is loaded with seed and super, and the rising roar of the engine sends it back into the air.

Our pilot for this exercise was Keith Wakeman and over the years that he did this work for us there were many amusing incidents. Keith had a reputation as rather a daredevil in the air but in fact, underneath his apparently happy-go-lucky exterior, he had a very shrewd and careful approach to flying. For him there was always humour to be found in any mishap. Once his driver rolled a smoke or performed some other such careless action and the truck ran off the road, turning the tractor upside-down in a shallow gully. 'Well,' said Keith, looking at it philosophically, 'I always wondered what a tractor looked like from below!' The loading arms which carried the bucket were badly bent so when we got the machine home we braced them against a sheep-yard post and hauled them straight again with another tractor. The bucket was rather lopsided after this treatment but it worked.

Another time Keith took Mary, who wanted to go to Darfield, in the plane as he was going home. The only paddock to land on there was behind the hospital, and he had to chase sheep off it with the plane before he could land.

'Are you supposed to land here?' Mary enquired rather anxiously.

'No,' said Keith, 'but I'll say I've got a pregnant woman who has got to get to the hospital quick if anyone asks me!'

Mary didn't think she looked pregnant, and certainly hoped she wasn't.

After the topdressing was safely over I took Keith on horseback to examine the hill he had been working over. His skill and competence in the air did not extend to what he considered the much more precarious control of a horse. I mounted him on the quietest hack and provided him with the largest stock saddle we possessed, fitted with nice big knee-pads for comfort and security. When he got back Mary asked him how he enjoyed his ride.

'Oh,' said Keith, 'it wasn't bad, but I wouldn't like to be up there without those stabilisers.'

Our agreement with the Catchment Board provided that no stock should be run on the block until the clover seed had germinated and grown enough to withstand some grazing, and even then only cattle should be used to start with.

The seed was sown in early July and there could be no hope of germination before September at the earliest. On the lower slopes clover alone was used but on the steep faces higher up where there was much bare ground some cocksfoot was added to the mixture. By early November it was clear that a lot of seed had struck and we were reasonably pleased with the result, though still very doubtful about its economics. The job had cost over £3000 ($6000) and that was a sum we could not have afforded without the pound for pound subsidy.

One year later there was a very different story. Heavy rain all through the spring of 1957 and January 1958 produced a growth of clover on the block beyond the wildest dreams of anyone concerned. Red clover reached 60 centimetres in height on Little Bailey and all the stock which I could concentrate did little to keep the feed down. Ewes with lambs at foot cannot be moved in any number; of wethers we did not have enough; and cattle—the obvious alternative—just died of bloat! Here was a tiger by the tail if you like; we had released a veritable atomic bomb and did not know how to control it. No need to worry on that score, however, Nature—that wily dame whose rules you break at your peril—produced from her capacious pocket an agent more powerful than any weapon at our disposal. In this deep mat of succulent grass the native porina moth—one of farmers' well known and constant enemies—laid many million eggs to hatch and steadily destroy clover and grass all through the winter, and when we examined the hill in the spring of 1958 we thought that everything we had sowed and much beside had been totally consumed. Again the dame turned on her compensation trick and six months later we found that enough clover seed had set during the boom period to reseed that block again, though with nothing like the incredible luxuriance of the year before.

So began our first real attempt to raise the productivity of tussock grassland. That we succeeded is shown by the stock figures. A dry, hard block which used to carry 700 wether hoggets for nine months now carries 1000 two-tooth ewes and sixty or seventy cattle, whose productivity is seven or or eight times greater. Not all our topdressing was as effective as this, but it opened the door to better performance from the whole flock and to the development of a substantial herd of cattle, for which Grasmere had a very limited range previously.

The phenomenal wet summer of 1957-8 not only affected the results of our topdressing experiment, it disrupted the whole of our summer work. Between mid-October and mid-January 90 centimetres of rain fell at Grasmere. What the fall was at Arthurs Pass I do not remember but it would be four or five times as much—nearly a year's rain in three months. During one of the peak floods the Waimakariri rose to 15 metres at the Gorge Bridge and every stream and river in the upper gorge was a raging torrent. Several times we had to rescue people from the Mount White road when they were trapped by the Hawdon or the Andrews streams. When we went to muster the Burnt Face for shearing one of our musterers, Ian Ellis, took his life in his hands and waded the Bruce Creek waist-deep to get on to the hill. His dogs were whirled away in the racing current and he was lucky that they were not drowned.

Haymaking and cultivation for winter feed were held up for

weeks on end and when at last the rain began to ease we had all the work to do in a brief period as the paddocks dried out. Our two boys, though only fourteen and sixteen, were roped in to drive tractors almost day and night and when at last the turnips were sown and the hay was in the sheds I thought they deserved a special reward before they went back to school. We decided to take horses and ride up the river to the Carrington Hut where the boys had never been. First we went up to our own Top Hut on Powers Country where the dry sheep were now grazing, partly to make sure they had not been driven down by the stormy weather and partly in the hope of getting a shot at some deer, for there were still plenty to be seen in the upper Waimakariri. We camped a night at the hut and saw no deer but we had a salutary lesson in the habits of another creature altogether. There was an old and rickety meat-safe which hung against the chimney at the back of the hut and in it we had placed our meat supply—enough chops to last the few nights of our projected trip. Alas, opossums broke open the flimsy catch and completely destroyed our precious chops. It was a surprise to me for I had never thought of the opossum as carnivorous. I thought they only ate leaves and grass and clover and fruit and vegetables from people's gardens!

There we were with very inadequate food supplies and several days to go. However, we were always hopeful of getting a deer to supplement the larder so we decided to carry on.

Now began the interesting part of our ride. The further we went up the river the more devastating was the erosion of the stream beds. We rode up the Jordan River in search of deer and were astonished to see that the bed of this stream had dropped as much as 3 metres in places. Where shingle had accumulated for many years, forming a fairly flat floor between the steep rock walls, there was now a clean line 3 metres above our feet to show what its level had been. Above this line ferns and shrubs clothed the banks; below it was freshly bared rock devoid of any growth. All this accumulated shingle and broken rock had been swept out into the main river.

The same picture greeted us at every tributary stream: the

Crow, the Anti-Crow, the Greenlaw and the White, all had been swept out as with a giant hose, washing a mountainous pile of loose rock and shingle into the Waimakariri. How far had it gone? How fast would it move? What effect would it have on our precious grazing flats at Cora Lynn and Grasmere? In the long run how much would build up behind Christchurch to swell the ever-rising mound upon which the river runs precariously above the City of the Plains?

Interesting thoughts indeed but they did not suffice to fill our bellies. We had shot without result at some swiftly moving chamois up the Jordan, but apart from those we had seen nothing, and a diet of bread and butter and porridge was getting monotonous; after a hungry night at the Carrington Hut we decided to ride home. At least, we said, there's always cold meat in the fridge at Grasmere. It was a long ride and, tired and excruciatingly hungry, we put our horses in the stable and raced inside to feast on the succulent mutton we had been tasting in anticipation for the last hour or two. To our horror and dismay the fridge, like Mother Hubbard's cupboard, was bare of the proverbial bone. Mary had gone away while we went on our little jaunt and it was not till she came home that we heard the explanation. She had met some of our young skiing and mountaineering friends who were going up to Arthurs Pass, and cheerfully invited them, if ever they were short of food, to drop in and help themselves. 'There's always plenty in the fridge,' said she in a burst of generous hospitality. 'Just help yourselves if there's nobody at home.'

Alas! They took her at her word, and several hungry young men enjoyed the last remnants of the feast which we had promised our aching stomachs!

In spite of our privations the whole trip had been intensely interesting and gave me much food for thought over the ensuing years. The flood waters where we had been exploring had come, except in the Jordan Valley, exclusively from country which had never been burned or stocked with sheep. If damaged vegetation had contributed to the erosion which occurred it must have been damaged by some other agency.

Deer sprang to mind as browsing in the bush and destroying the subalpine shrubs, and above the bushline there were chamois in quite substantial numbers. Undoubtedly if these creatures were responsible for any part of the massive outflow of shingle into the main river, then the authorities were right to make the Waimakariri Valley a top priority area for noxious animal destruction.

Chapter 18

IT TAKES ALL SORTS

> I thought that kitten must be a Persian—it purrs a lot anyway!
>
> KEITH THE COWMAN

IN MANY YEARS of station management a large number of employees come and go. Some one remembers with affection, some with amusement, and a few with downright horror. One man whom we will always remember with affection was Tom Ropu. Tom first came, as I recall, in the early fifties when the weaning muster spread over some weeks, during which mobs of ewes and lambs came in and the lambs we intended to keep were crutched and dipped in comparatively small numbers. One crutcher was sufficient for this sort of job, and a man was needed who would look after himself—fill his own catching pen and sweep the board from time to time himself. Tom was perfectly happy to do this; time was of no great concern to him and his conversation was brief and laconic, almost a caricature of the stage Maori who has little command of English. He was born in the Chatham Islands, and probably never spoke English until he came to the mainland. I once asked him if he ever went back there.

'I go once,' he said, grinning broadly. 'No bloody good! Old days, everyone go to the pub—stay there for a week. Now, all got motorcars—stay one day and go home. Chatham Islands no bloody good any more!'

One of Tom's great virtues was that he didn't much care if we had a cook or not; any catering difficulty was easily overcome because all he really wanted was a leg of mutton and a cup of tea. Once when the cook had disappeared or failed to come home I found Tom crutching away in the shed when I came down after breakfast.

'Did you get some breakfast, Tom?' I asked. 'Yes, Boss,' he said. 'I had the leg of mutton we had for tea last night.' And there beside him on the window sill was the clean-picked bone from which he had made his breakfast while he worked!

For several years Tom came by train when he was needed, but a day came when he turned up driving a little, covered van. He had a grin all over his dusky face and I regret to say that he was waving an empty beer bottle out of the window by way of greeting. The van was his joy—I wouldn't say his pride because he had none—and it was not long before it had numerous marks of ill-treatment on it. He described with vivid gestures how he had failed to take a bend on the Motunau road one day—perhaps a beer bottle had something to do with this mishap. The van had swerved off the road and down a bank, ending its progress upside-down in a deep ditch. Tom illustrated how he slowly recovered consciousness; how he slapped his body and each of his legs; and how he discovered to his delight, 'I not dead—I all alive still!' What's more the van was all alive still also, if somewhat dented on the top.

He never really discovered how it worked, and if the slightest thing went wrong someone would have to go and fix it for him.

Tom didn't like to have his wages forms filled in and the tax deducted. He always claimed that he paid his tax on his earnings at the end of the year, and I know that the Inland Revenue were always trying to catch up with him. His English may have been sketchy but his logic was far from it: 'If I pay tax now the money gone; if I wait till end of year I might die—pay no tax at all!' And strange to say that is just what happened. Tom died on the shearing board at the Terrace Station, victim perhaps of too much meat and beer, and I expect the Inland Revenue went short of its pound of flesh, as Tom always hoped it would.

Cowman-gardeners have always provided a number of curious types. Many were transitory phenomena who passed across our lives like comets across the heavens, trailing faint clouds of reminiscence and leaving little mark on the life of the station. Percy was one of the more permanent of these men, and though in many ways he was far from an admirable character, we still remember him with some affection. He was a Yorkshireman who came to New Zealand after World War I in which he served briefly in an ill-fated expedition which went to Archangel during the Bolshevik revolution. There was no fighting there, and Percy's part seemed to have been divided between trying to steal food, which was scarce and unpalatable, and carrying buckets of water which, he assured us, would freeze while you were carrying them if you weren't very quick. Percy was a product of the worst days of industrial England, when half-grown boys went down the coal pits with bones and joints already weakened by malnutrition, and faces paled with lack of sun under the ingrained grime and coal dust. He also had much of the brutal selfishness which is engendered by such conditions, and the dour, humourless wit of the Yorkshireman anywhere.

He could be quite ruthless if he got the chance. He had a constant war of nerves with any cook whom we might have and if a new one arrived he took steps as soon as possible to assert his paramount position. If the cook wanted to get milk and vegetables delivered when they were needed he or she had better make sure they made a fuss of Percy. To one he actually said on the first day: 'Better be careful how you treat me. I cut a lot of ice round here.'

What he said in the town about his position of importance we never suspected until one day when I took a photo of him with a cabbage he had grown. It was truly enormous and he loved being photographed anyway.

'That's me with me broossells sprout,' he said delightedly when he saw it.

Unfortunately the picture was so good that I sent it to the *Auckland Weekly* magazine—one of the old illustrated papers, now defunct. Percy was horrified. 'Yer shouldn't 'ave doon that,' he said indignantly. 'All me cobbers in town think I'm the manager up here—now they'll know I'm only the cowman!'

One of Percy's prize possessions was his false teeth. They were very well preserved, because in fact he never used them for the purpose for which they were intended; they were strictly for ornament and he never put them in unless he knew somebody was going to take his photo, or when he was dressed in his suit and off to town. Then he appeared with an unnatural smile which gave his face a look of spurious bonhomie quite unlike his usual dour expression.

After one of his brief holidays in Christchurch he came home with a worried and anxious look—and without his teeth. He asked if we would mind ringing up a tea shop in Cathedral Square. 'Yer see,' he said, 'I went in there to have a cup o' tea and took me teeth out to eat a pie and left them on the winder sill.' We pictured the waitress's face when she found them grinning at her, but perhaps this is an occupational hazard for waitresses; at any rate she kept them for him and the management kindly sent them back.

Percy had the Yorkshire passion for cricket and when Len Hutton brought a team to Christchurch I had to let him off to go and watch. When he came back he had a grin all over his face and he couldn't wait to tell us about his triumph.

'I saw Mrs 'ootton,' he said delightedly. (It was nothing to him that she was already Lady Hutton.)

'Did you speak to her, Percy?'

'Aye, I spoke to her.'

'What did you say to her?'

'Yer woon't git no Yorkshire puddin' 'ere, Moom!'

'What did she say, Perce?'

'She said, "Woon't I?"'

A conversational coup long remembered and repeated.

One of Percy's great joys was to be invited in for a drink when our son-in-law, Archie Mackenzie, was with us. With a good, stiff whisky in his hand he made the most of half an hour's chat; he would surreptitiously fill his glass again when he thought nobody was looking, and his accent would get broader and broader and his reminiscences longer.

Archie often used to bring a tape-recorder and spend a morning taping medical records for his secretary to type. One day he put the recorder on the kitchen table, screened by a box, and invited Percy for his regular session. After the second drink he began to attack me about a cow I had bought.

'Yer didn't oughter go buying cows without takin' me,' he said. 'Look at that cow with the big tits—I mean teats,' he corrected himself hastily, in deference to the ladies present. 'She ain't no good and never will be. I could ha' told yer that as soon as I saw her.'

The diatribe went on for some time, and I'm afraid the family had a good laugh at Percy's expense when Archie played the tape back later. Perhaps it sounds an unkind trick, but Archie was to reward Perce in full measure. His poor old knees, weakened by rickets and malnutrition, became so bent in the end that the only way he could carry a milk-bucket was to swing it in the bow between his legs. Walking from the cowshed to the vegetable garden became an agony to him. Archie got him into hospital and operated on both his knees and, after a long convalescence, during which he was inclined to moan because he wasn't cured overnight, he was able in the end to walk comparatively straight and without a stick. To anyone who saw him before, it was indeed a miracle. There was one good thing about his long session among the nurses—he had to wash!

One of the most wonderful men who worked for us was Jimmie Mitchell, an Australian shearer who first came to us in 1954. It was remarkable enough that he was then seventy-six years old, and still able to shear a creditable number of sheep—over 100 ewes a day and eighty to ninety wethers. More remarkable still, he told us how in 1904, fifty years before, he had been engaged as a shearer at Castlehill, halfway between Springfield and Grasmere. The coach driver would not carry his swag and so he had to hump it and walk the 30 kilometres to Castlehill. There he shore in the old woolshed down by the Porter River, and when the shearing there was finished he carried his swag again another 25 kilometres to Grasmere, where he shore again, finally going on to Cora Lynn, which was then a separate station owned by the McKays. So fifty years later he was back, with his mind alert and his memory clear about things that happened there. When you consider how many sheds he must have visited in the intervening years it is astonishing that he should remember the jokes and horseplay and the location of buildings after all that time. Jimmie shore for us for three more years, till he was seventy-nine, and didn't stop shearing then, but I don't think he came to New Zealand after that. I do know though that at the age of ninety he did a demonstration of blade shearing for Australian television! A marvellous and charming old man.

After Percy's departure we had for several years one of those unfortunates who often drifted into the job of cowman-gardener because their mental processes were below the average and the speed required to dig a garden and milk a cow marked the limit of their capabilities. Even in such simple tasks Bob was sometimes confronted with a situation beyond his capacity—often, I fear, as the result of practical jokes. In our old cow-bail, as in so many others, the cows walked in from the back and put their heads between two vertical bars, one of which was then pushed across and pegged by the cowman. To do this he had to walk up beside the cow and she could, of course, back out before he could close the bail. To avoid this I had arranged two cords which could be worked from behind—one to close the bail which then locked with a drop block, and the other to remove the block and free the cow when she was finished. Very simple and efficient, and well within Bob's mental grasp.

Unfortunately, young musterers are often short of some innocent amusement when they are not working. The musterers' hut being in full view of the cowshed, a couple of the bright lads thought it would be fun to reverse the cords and watch to see what happened one Sunday afternoon at milking time. The trick worked like a charm; no sooner did the cow enter the bail than she came whizzing out again when the wrong cord was pulled. Bob scratched his head and tried again and yet again. No matter which cow he tried, the same thing happened until he was in despair. Finally he gave it up and went up to the cookshop to seek advice from the cook, who was a friendly soul. Friendly or not he was quite ready to enter

IT TAKES ALL SORTS

into the spirit of the joke, and he told Bob with all seriousness that it was the weather; exactly the same thing had happened when he was cooking at Stoneyhurst when the wind was in a certain quarter. 'Yes,' he said, poking his head out of the kitchen door, 'the wind's blowing just the same way today!'

In the end, Bob was reduced to bringing his dilemma to the boss and I was solemnly led up to the cow-bail to try to solve the mystery. Needless to say it didn't take me long, but when I came out and looked up at the musterers' hut the grinning faces had all discreetly disappeared.

The cook, as I have said, was not averse to pulling Bob's leg. He stayed with us only briefly but he was a very good cook. There is a system, he told us, of 'delivery crews' in the ship-building ports of northern England—men who take a new-built ship across the world to deliver to its purchaser and then go home. Crews must have cooks, and their training for these voyages is well organised and efficient. George had come to New Zealand on a new dredge which had been built for the Timaru Harbour Board, and for a few months he saw the country by cooking on sheep stations. While he was with us another plot was hatched in which he and the other men all took part.

The kitchen is at one end of the long shearers' quarters and the cook's room opens off it so that he is separated from the shearers by the big dining room. When we were not shearing, the station hands usually had their meals at the kitchen table. One day at breakfast time George poured out a cup of tea and made an extra piece of toast and carried it into his bedroom, carefully closing the door so that the watchers could not see into the room. Bob observed this with some surprise, particularly when George's voice could be heard apparently talking to someone. When he came out he again carefully manoeuvred the door so that not a glimpse of the room could be seen. When the men had finished and were filing out George went in again and came out with an empty cup and plate. Bob was fairly bursting with curiosity, and as soon as they were all outside he asked one of the musterers who on earth George had in his room.

'Oh,' said the conspirator, 'didn't you know? George has got a girl in there. She came up on the train last night and slept in George's room.'

The game went on at midday, when George served a tasty plateful of meat and potatoes and lettuce and took it to his mythical paramour. By the time the meal was over Bob was determined to get a look at this voluptuous tart—for the other men had described her charms in glowing, if somewhat indecent language. He tried the window of the cook's room, but it was tightly closed and the curtains drawn. He snooped and listened and found excuses to loiter near the cookshop at frequent intervals, bringing milk and cream and unnecessary vegetables in the hope of finding George absent for long enough to let him get inside, but George was far too shrewd and was getting endless amusement from watching these antics. Bob was consumed by jealousy and told the others that it wasn't fair—why should George be allowed a girl when he wasn't? 'Ah,' they said, 'cooks are hard to get these days, and the boss wouldn't dare say anything to George, or he might leave!'

In the end, of course, the lady had to depart the way she came, in the dead of night. Morning revealed only an open door, and an empty bed in George's room.

I have already mentioned one or two of the weird variety of men who came to us as cooks, but there were many others, some comic and some very tragic.

Tom was a slim, gentle man of indeterminate age. His manners were delightful and he was a good, clean cook. His only failing was that he could cope with cooking for only a few men. Nature had not endowed him with quickness or the gift of organisation. He insisted on feeding the men in the large dining room because he did not like the kitchen cluttered with their noisy, dirty presence and for the same reason would not let them come and fetch their own food. He would carry it to them one plate at a time, forget some important item, like the potatoes on somebody's plate, and have to take it back and make the journey again. In the course of a day he must have walked for many a weary mile and it's no wonder that by evening he was worn out. When he managed to find a few minutes of leisure during his self-inflicted labours, he had one supreme method of relaxation, and I must admit that I never saw a man look as if he enjoyed his rest period more. If he had stretched himself on a rug in the sunshine and slept on the shores of a tropical island he could hardly have given a more convincing picture of total satisfaction.

Tom had a car—an old car with a battered hood and open sides; somewhat uncertain of temperament and not suitable for serious motoring. Tom didn't mind—he only wanted it to sit in. There he would be, lolling in the front seat, a cigarette in his idle fingers raised occasionally to his lips to send out a puff of smoke which dissipated in the balmy air; a picture of perfect contentment. He didn't have the worry of driving the thing, or even the anxiety of seeing whether it would start; it was his, and he could sit in its worn but enchanting upholstery secure in the knowledge that he was a man of property.

But there was a sad day in store for poor Tom; one day he did decide to go for a drive, and, of course, the car would not start. He pressed the starter, he wound the handle, he pushed it down the little hill, still it wouldn't go. In the end he got the musterers to push it for him. Down the drive they went, Tom sitting anxiously in the driver's seat, bolt upright and pressing with his hands and feet as if to force the engine into life, the crouching figures of the men leaning against its battered stern and kicking up the gravel with their heels as they raced it down the road. It was all to no avail, and with the motive power exhausted the car came to rest at the front paddock gate. Tom had to cook the dinner and the men had other things to do, so there they left it, abandoned and forlorn.

During the night a nor'-west wind got up; it howled round the house lifting loose paper, sheets of tin and boxes and sending them careering across the paddocks to finish up against a fence. Some time during the night one of its whirling gusts caught the hood of poor Tom's car, wrenched apart its ancient stitching and tore it clean adrift. There she sat, bent framework and bare sky all that was left of a once glorious lid. Tom spotted it at breakfast time, and we saw his tall, angular figure flying down the paddock before the still frolicsome wind, his arms uplifted in despair, and crying plaintively, 'The car, the car! Oh! Oh! the car!'

Tom went with us on one of our trips to the Top Hut in the autumn muster. Getting him prepared to go was a stupendous operation. The organisation of all that gear, the lists of stores required, the calculation of quantities, the packing of the plates and cooking utensils, the anxious desire to take everything that might be useful, the pre-cooking of cakes and duffs and cold meat; all was a source of worry and anxiety.

Just as we were ready to start—the men and horses had gone on, and Mary, as usual, was left to bring the cook—poor Tom collapsed. He announced almost with tears in his eyes that he hadn't had time to have a cup of tea for a fortnight and was completely exhausted. Tact and the blarney finally got him on the way, and once at the hut the whole business became quite a picnic to him. He made endless cups of tea for us and cut our lunches, and when we came back at the end of each day there was hot food and a fresh-boiled billy, and Tom's lively figure bobbing in and out of the hut door. At least he couldn't carry plates back and forth because the hut is so small you are never more than arm's length from anyone.

The most tragic of our cooks was Mrs D. Perhaps hers is the most poignant story of them all because, unlike most of the human derelicts who came our way, Mrs D.'s tragedy involved a child. They came to us when Dawn was twelve years old, a child who should have had the benefit of a normal school education among children of her own age, but who had been dragged around the country as her mother drifted from job to job. Sometimes they lived in pubs where the mother cooked; sometimes with travelling shearing gangs; sometimes for a longer spell on stations like our own. The girl was never at a school long enough to make any real progress, and in between whiles the Correspondence School coped as best it could with half-completed lessons sent from ever-changing addresses. When Dawn was at Grasmere Mary tried to supervise her lessons, but found the poor child woefully ignorant and difficult to teach.

When she was properly on the job Mrs D. was a capable and efficient cook, but the demon had her by the throat, and every few weeks she had to go away to 'sign some paper' or 'complete some legal business'. Dawn was convinced that she would be well off at some vague day in the future, when the property her mother really owned would come to hand. She was brought up to believe that she came of some good family, and to prove it Mrs D. used to describe how she had inherited Lord Nelson's telescope. We never saw this famous article, but poor little Dawn was convinced that it was the very one which he placed to his blind eye at Copenhagen.

When she had earned enough Mrs D. would take off for Christchurch, dragging Dawn with her. There, while the money lasted, they travelled in taxis and drove from pub to pub and shop to shop so that she could live for a few short days in the style to which she believed she was born. Then, when the orgy of spending ended, she would come back, bringing expensive foods like hams and ducks to show the station hands how she would feed them if she had her way. On one of these occasions she gave a party in the cookshop for Dawn's thirteenth birthday. Everyone on the place was invited, including Mary and me. There was to be a magnificent supper, and champagne to wash it down, and everybody was to be dressed in their best for the occasion. Mrs D. was to wear a long white frock—only unfortunately she found it difficult to put it on after trying out the champagne to see that it was good. One of the men came into the kitchen to find her prostrate on the floor, entangled in her flowing robe, which Dawn had been trying to fasten up the back. There she lay, struggling like a cast sheep and crying, 'Pick your mother up, Dawn. Pick your mother up!'

There was worse to come—Mrs D. became very ill. It was not surprising, perhaps, that her constitution could not stand the life she led. She was taken into Christchurch Hospital for tests, and we kept Dawn at home, wondering what we could do with the child if Mrs D. did not come out.

Finally, after about a week, she was told she could leave the hospital, but was not told the result of the tests. She wanted to come back to us, but she looked so ill we were sure she was not fit to work; we felt we had to find out what was really wrong with her. The hospital would tell us nothing, but we had ways of enquiring and were horrified to learn that she had cancer of the liver and her life could be measured in months. It seemed to us very wrong that she had been sent out to put in the time as best she could until the inevitable day when she would go back to die.

We didn't know what to do; Grasmere was no place for a slowly dying woman. We would have to find somewhere for her to go.

Religion played but little part in Mrs D.'s life no doubt, but it was to the Presbyterian Church we turned for help, and we found it in full measure. The hospital visitor was Arthur Mitchell, who had conducted the marriage service at Arthurs Pass chapel when Archie Mackenzie and our daughter Catherine were married. He found a flat where Mrs D. could live, and persuaded her that she could not go back to Grasmere—for she was convinced that she could still do her job. The flat was furnished, but there were no cooking utensils or cutlery, so when I brought Dawn down from Grasmere I had to go shopping to provide these necessities. Now for a brief period the comic side of this sad story intervened. It was the day of the New Zealand Cup at Riccarton, the city was packed with cars, and parking was impossible. I had to leave the car somewhere while Dawn and I went shopping, and I could find nowhere but the sedate premises of the Christchurch Club, whose members are always careful to appear respectable. One of Mrs D.'s peculiarities was to dress her daughter in the fashion of a long past generation. In her best clothes she looked exactly like John Tenniel's illustrations for *Alice in Wonderland*. The long hair kept in place by a ribbon; the long dress; even the little pinafore was there, and the effect was sufficiently unusual to attract the startled gaze of anybody in the modern world. I knew when I went in there that I was sure to meet somebody who would recognise me, and sure enough, no sooner had I disembarked with my strange protégée than we came face to face with a member I knew well. The novel *Lolita* was not many years published, and the man in question could have been forgiven for thinking that I was about to repeat that disreputable story in the respectable world of Canterbury farmers. Needless to say, I hustled the child out of there as quickly as possible; but even the business of shopping was almost equally embarrassing, because I felt sure everyone who served us pictured an illicit *ménage* which a lecherous old man was setting up with this innocent-looking virgin.

The flat was not needed for very long. Mrs D. returned to the hospital in a matter of weeks, and Dawn was left a hapless orphan. Here again Arthur Mitchell became the fairy godfather for he managed to persuade a reluctant uncle to provide the poor child with a home and schooling.

Such an upbringing is an impossible handicap, and in spite of one or two satisfactory jobs after she left school, adult life landed Dawn with an illegitimate child in the concrete jungle of an Australian city. The last episode in this pathetic story was an illiterate letter addressed to us from the father of this child, a steward on a P. & O. ship, begging for Dawn's address so that he could go back and see his son.

Alas, we couldn't give it to him.

Chapter 19

THE GRASMERE DOG TRIALS

> 'What happened to that last mob of sheep I hunted down to you, Boss?'
> 'I chased them ahead on to tomorrow's beat.'
> 'Ah! but yer didn't get 'em in the yard, Boss. Yer didn't get 'em in the f—ing yard!'
> OLD MUSTERER TO C. C. BURDON

IN THE EARLY 1950s we held a competition which has become an established institution in the Waimakariri Gorge. I was asked to judge a dog trial which had become an annual event at Mount Possession in the Ashburton Gorge. For the benefit and entertainment of the local musterers, Sam Chaffey, whom I have mentioned as figuring in the activities of the Guide Platoon during the war, organised a three-dog event in which the shepherd had to work five sheep with three consecutive dogs across a rugged little mountain face above the Hakatere Station. The competition was confined to men mustering in the gorge, and judging was based on the successful performance of the task—i.e. getting the five sheep to the end of the course, rather than on the technical niceties of a normal dog trial. I enjoyed the day's judging except for the fact that the weather was threatening and sometimes both dog and sheep were hidden by swirling mist. The course was fairly tough, particularly for the first section, which was quite a long 'head', and the last, where a huntaway dog had to stop the sheep from disappearing round the hill and turn them down through rugged rocks to the finish. Quite a difficult job requiring a very 'handy' type of dog. I only had one problem during the day, and that was when I suddenly noticed that a competitor had started with six sheep instead of five. Should I try to stop him? Very difficult because I was a long way off and there was none too much time to give him another run; so I decided to let him go and see what happened. He then lost one sheep. Dilemma again; he still had his proper quota! I let him go on again, and to my considerable relief he lost the rest in the rocks at the start of the second leg so I had no awkward decision to make.

The whole affair impressed me as a first class competition for men who made their living by doing exactly that kind of thing, and knew full well that if they could not do it successfully they ought to be able to. The reward was to get your name on the silver replica of a mustering billy, the old-fashioned wide-bottomed pot with a mug which fitted inside to form a lid, and was boiled, when time permitted, hanging from the end of a mustering stick over a hastily-lit fire. One of my old mustering companions Hugh Knubley had one which he called tin-arse, because it was always the last to boil.

I came home determined to try to institute a similar event at Grasmere and in 1953 or 1954 we did so.

Designing a course was an interesting exercise, and ours turned out to be quite different from the Hakatere one. Behind the Grasmere homestead is a small conical peak called Little Bailey—an offshoot, as it were, of the big hill above it which is really the north end of the Craigieburn Range. Half way up it there is a convenient little outcrop of rock, too small to be called a bluff but making a good vantage point. Here we decided the competitor would stand and send his heading dog to catch five sheep released from a pen just out of sight round the curve of the hill below him. The pull was not so long as the Hakatere one but it proved to be more difficult for several reasons. One was the steepness of the hill, and another a strip of dense, high tussock which marked a tiny, swampy gully. There were all sorts of ways of getting into trouble and the dog which got full points did very well indeed. The man had then to pass the sheep above the rocks and take control of them with his second dog, and many a man came unstuck in that tricky exercise. He had to turn them downhill and his dog must shepherd them along below two parallel stakes and then lift them up above a third, while the man walked round the hill above. Did I say walked? Mostly he had to run, stumbling through rocks and tussocks to keep his sheep and dog in sight because the hill is convex and his quarry constantly disappeared like ships dropping below the horizon at sea. If he had luck and a good handy dog he would arrive triumphant at the top of the last section, where a smooth, clear tussock face leads straight downhill to the finish.

This looks deceptively easy to the watchers down below; it's not very far. But sheep do not like going straight down a steep hill, and given the slightest chance they will angle off across the slope, losing points for a dog who cannot keep them straight and sometimes even missing the wide-spaced finishing poles altogether while the man above dances and screams in impotent rage. It's all great fun for the watchers, and luckily the hill is so close to home that the runs can be watched throughout from a car on the flat or a sheepyard rail—even from the door of the well patronised bar in the blacksmith's shop. Besides the hill event we have a trial for handy dogs in which one dog must take three sheep through a set course, heading and hunting them as required. This keeps the competitors occupied before and after they have their runs on the hill and keeps the bar from being over-patronised, which can at times occur!

We always had the trial on Easter Sunday because the stations are not usually mustering at Easter and the back-country man's chief relaxation, deer shooting, can be very dangerous when the 'sportsmen' from the city are at large. Bullets are often flung at anything that moves and the men are better occupied where they are safe. To finance some fairly generous prizes, besides the silver billy which we cribbed from Hakatere because we couldn't think of any better trophy, the stations which participate levy themselves on a per-sheep basis to provide extra funds, and the whole thing becomes a community effort in which the whole gorge comes together for one day in the year. Not only the owners, managers and shepherds are involved but anyone working on the stations makes it a day out. Our old friend Percy looked forward to it with pleasure because he always kept the bar. Although he made the best possible use of his position he never allowed anyone else to get away with anything. The cost of the beer had to be recovered before Percy would let anybody have a free drink, and we could be sure that there would be a surplus in the bar account as long as he was in charge.

There were times, though, when Percy's pleasure was somewhat marred by the rude behaviour of the young musterers. Unfortunately for him he always had to leave his post at a time when the lads were beginning to get hilarious, and since the cowshed was close to the bar, it was an obvious target for a bit of horse-play—or should I say cow-play? Riding Percy's best cow was one of the favourite tricks, and a well-fed cow can be quite a lively mount so the lads often got a dumping in the mud for their pains. However, this treatment did not exactly pacify the cows and there was apt to be a lot of trouble when Percy went to get his scattered herd assembled and, one by one, into the bail. There was one terrible day when his favourite white cow would not go into the bail after her treatment by the boys, and the enraged Percy took a large stick and beat her over the rump. The cow thereupon took the only revenge she could and lifting her tail sprayed the unfortunate Percy from head to foot with a stream of liquid ordure. Naturally the practical jokers were highly delighted with this appropriate retribution and went back to the bar to celebrate it. They were not quite so delighted when they were offered a revolting-smelling khaki liquid instead of milk for their cup of tea after the trials were over. Needless to say nobody drank milk that night! After the judging was finished and the prizes distributed with due ceremony and much clapping of bashful winners, most of the competitors and some of the spectators went to the Bealey Hotel. Percy was a passenger in one of the vehicles and it was soon discovered that, though he had changed his shirt and put on a jacket, his boots and trousers were still those which he had had on in the cowshed. It's a wonder that they didn't put him out on the road, but he certainly had one end of the bar all to himself when he got to the pub.

One of the great beauties of Grasmere is the lake. It lies against the steep flank of the Long Hill where all the surface water from the hills behind the homestead comes up in never-failing springs which keep it at a relatively stable level. The strip of mountain beech forest on the hill behind throws a deep green shadow on the water when the sun is low and the surface still, and the only disturbance of its dusky mirror is the splashing of innumerable birds and the occasional ring of a rising fish. Mary and I used to delight, when we had an hour or two to spare, in sailing or rowing the small boat we kept in one of the tiny harbours on the shallow shore where our paddocks came down to the water's edge. These harbours were really spring holes where you could see the bubbles rising from the incoming water and quite deep little channels led out into the main lake. The channels, like most natural things, were not straight and to sail in or out of the boat harbour without the centre board grounding in the mud was a highly skilled operation for which one needed a local pilot's licence. When we could not sail we often took a thermos of tea or a picnic lunch and rowed about and watched the birds.

Grasmere is one of the few lakes where the crested grebe still nests and this is largely due to the fact that it is almost impossible to approach the water from the Long Hill side and so the nests are undisturbed. Once when we were drifting slowly along the shore with a scarcely breathing wind we found a nest in which one egg remained. Others had hatched, as we could see by the broken shells, so we took the last, abandoned egg and sailed on home. To our surprise we heard a sound from the egg and when we put it to our ears we could hear that there was a live bird inside. It was no use taking it back, for the nest had clearly been abandoned. We finally broke the egg and extracted a tiny, weak, struggling chick, which we placed in a cardboard box on the top of the stove. It livened up with the warmth, and cheeped mournfully at the strange world in which it found itself. What, we wondered, does a grebe chick eat? All we could think of was cooked spinach as a substitute for the lake weeds which probably would have sustained it in its natural state.

I don't suppose it was strong enough to survive in any case, but spinach was obviously not the right diet for it soon died.

One day in the mid 1950s we were spending a peaceful Sunday afternoon paddling quietly about on the glassy lake. The sun shone down from a cloudless sky, picking out in almost dazzling colours the different birds which, like ourselves, were revelling in the peace and beauty of the scene. There were many paradise ducks, little flocks of greys and mallard, black teal, black swans, two or three pairs of grebe and, as always, a lot of Canada geese. Suddenly there was a bang which reverberated round the whole valley, echoing off the steep northern shore of the lake. Close behind us a paradise drake keeled over on the water and a crimson stain spread slowly round it. The dark head, which had been plunging under the surface, feeding, lay flat with its beak open, and the gay feathers of its underwings were half spread out. A feeling of intense anger swept over us both. The shattered peace, the wanton killing, the destruction of something beautiful, and the pathetic remains drifting lonely on the water—for the other birds had fled hastily at the shot. All these, and perhaps resentment that *our* lake, under our very noses, was the scene, combined to swell this sudden feeling of outrage. Only one more thing was needed to cause us to give expression to our fury and this we got. A man's figure rose from the tall grass of the lake shore and shouted to us across the water, 'Hey! Pick up that duck for us will yer!' It was enough. I shouted back, 'NO!' and rowed away.

When we got home I wrote to the Minister of Internal Affairs suggesting that Lake Grasmere should be made a Wild Life Refuge, and after some time this was done in 1957. It not only prevented the kind of wanton destruction which we had seen but, since power boats were banned from the refuge, it removed a threat to the grebes' nests which floated on the water or lodged just above it on the far bank. If power boats and water skiing had been allowed on the lake that would have been the end of the grebe.

The scheme was not without its problems, on account of the Canada geese, but I have never for a moment regretted the action that we took, and I have often deplored the poor sportsmanship of certain members of the North Canterbury Acclimatisation Society who have sought to have the dedication lifted to give them access to more shooting.

Canada geese have always been a problem in the Waimakariri Valley. When I came to Grasmere there were often about two hundred feeding in the paddocks adjoining the lake. Jim Milliken had the same trouble at Flock Hill, where his cultivated ground ran down to Lake Pearson. After some experiments he succeeded in poisoning a large number and for many years after that neither he nor I suffered much from their depredations. However, they gradually returned and by the 1950s were again a serious menace to crops, in ever-increasing numbers. Once the Wild Life Refuge was established we began to call upon the Acclimatisation Society to remove the geese when they became a menace, which soon became a regular event. The society rangers were very helpful and did what they could by placing automatic scare guns in the paddocks near the lake. These, of course, are of rather limited effect because they destroy no geese, and if they are placed at Grasmere the geese go to Flock Hill, and vice versa.

In conjunction with the guns we constructed a life-like scarecrow, with body, arms and legs constructed of sacks stuffed with hay, surmounted by a large hat and armed with a wooden 'gun'. This was tied to a stake not far from one of the scare guns. Its physical appearance was sufficiently life-like, we hoped, to repel a goose, but its arms and legs were not quite as nature usually constructs them. The legs particularly were apt to bend if the figure sagged on its supporting post, and in consequence it was promptly christened Percy. 'Percy' survived for quite a while and when not in active use he was brought home and stored, because the hay in his interior was a ready target for any cattle in the paddocks who, unlike the geese, did not regard him with suspicion. After he had been disembowelled several times we thought it better to bring him home. This led to another series of practical jokes, the first of them played on poor Bob, who had succeeded Percy in the cowman's job. The other men found some red paint which they plastered realistically down 'Percy's' ample bosom; they put a rope round his neck and hung him from the veranda in front of Bob's hut. He looked horribly dead, and the shock to Bob's system when he found the corpse in the half light of early morning, on going to fetch his cows, can be well imagined. The men were so pleased with the result of this grisly joke that they put 'Percy' on the seat of the cook's lavatory for the unfortunate woman to find when she got up in the morning. Luckily, she was made of stronger stuff than Bob, and pitched 'Percy' out of the door neck and crop.

Employees were not the only amusing and interesting characters we had at Grasmere. In 1956 a delegation of Russian farm officials came to New Zealand and, thinking that high-country farming might have parallels in Russia, the Department of Agriculture asked if we would have them to look at our grazing system. They arrived in big, black, official cars, and they looked like big, black, official caricatures of Russians: big overcoats, hats pulled down over large, stolid faces, hands buried in the pockets of their coats. They disembarked and stood in a crowded group on the narrow veranda of the house and I greeted the nearest with an out-stretched hand, only to find that I had made a frightful gaffe and caused acute embarrassment to the man concerned. There was a high official present—none other than the Deputy Commissar for Agriculture of the Union of Soviet Socialist Republics—and I had failed to greet his lordship first!

In spite of my initial blunder the rest of the delegation proved to be reasonably human, and once we got out into the open and began looking at sheep and grasses we found a common understanding. One thing stood out—although they had an excellent interpreter who had been at Cambridge, the value of Latin as an international means of communication in science became apparent. We all knew the Latin names for plants and grasses, though pronunciation was often very different. The visit ended with some afternoon tea, for which Mary had baked her usual excellent cakes. We wondered if the Russian menu does not run to cakes because we never saw it vanish with quite such speed. In about two minutes there was nothing left! The only man who let his concentration wander from the feast was the Commissar, and he, much to our amusement, found and became absorbed in a copy of *Esquire*—a mild enough journal in these days, but in 1956 it was salacious enough to interest males of any race. Pictures are a lingua franca which transcend the limitations of the tongue, and no doubt such entertainment was rare in the Soviet Union in those days. At any rate the Commissar did not offer to share it with his subordinates!

Another visitor from the same source—for the Department of Agriculture, like other institutions, found Grasmere a convenient and attractive place to take visitors—was a South American. I think he came from Brazil but I cannot now remember. He had a charming personality and the romantic name of Lohengrin Gonçalvez. We spent a pleasant afternoon looking at Grasmere and Cora Lynn, and found him well informed about sheep farming. The language difficulty appeared again on several occasions and I remember trying to explain to him the external parasites of sheep for which we were dipping. I referred to lice—though in fact we never had such things at Grasmere—and his puzzled expression showed that he didn't know the word. Suddenly his face changed and a broad smile spread across it. 'Ah!' he cried delightedly, 'I know, I know: louse—lice; Micky Mouse—Micky Mice!' Another international language had come unexpectedly to the rescue.

One of the most charming visitors we ever had at Grasmere was Mingma the Sherpa. In 1961 Sir Edmund Hillary led an expedition to the Himalayas to climb Mount Makalu and with it went our son-in-law John Harrison. It was an unusual expedition in several respects and it turned out to be an ill-fated one.

The plan was to climb the mountain without the help of oxygen; it had already been climbed by a French team with that vital aid, and in order to see whether Europeans could acclimatise to oxygen shortage by living at a high altitude, a large party spent the previous winter in a hut at 6000 metres. John Harrison, who had recently married our daughter Anne, was to be one of the later additions to the team who went straight from a low to a high altitude without preliminary acclimatisation other than the march through the hills.

It appeared in due course that this was no disadvantage, but long before it was put to the test the expedition suffered a mortal blow when Hillary himself was affected by a kind of stroke and had to be taken down to a low altitude to recover. While he, poor man, had to endure the frustration of complete inactivity—surely the greatest punishment which could be inflicted on such a man—the expedition went on without him.

Bad weather, injuries and frostbite dogged their attempt to reach the top, and finally the whole team had to withdraw. Peter Mulgrew, who lost his legs from frostbite, summed it up in the title of his book, *No Place for Men*.

All this seems a far cry from Grasmere, but a few years after the expedition Ed Hillary brought to New Zealand his leading Sherpa, Mingma Tsering, a very just reward for the devotion and courage these wonderful little men displayed. John brought Mingma to stay with us at Grasmere, the sort of mountain country which might make him feel at home, and in spite of the primitive life which he had led we found he had already adapted himself to European customs—knives and forks, pyjamas, collars and ties; he used them all as if he had been brought up with them and his natural good manners were a delight. One or two little remarks betrayed his unfamiliarity with our world. He said to me one day, with a note of respect in his voice, 'John Harrison very rich man, he have two cameras and a car!' John was then a young commercial artist with few assets beyond a well-worn Volkswagen to get him to his beloved mountains; but perhaps Mingma was right and he *was* a wealthy man in the things that really matter.

The Romans had a saying, 'Those whom the gods love die young.' The gods of the mountains must have loved John, for he died among them, as he would have wished, leaving great grief among us all.

The thing which impressed Mingma more than anything else at Grasmere was that very primitive instrument, a pair of blade shears. He had been telling us how the Sherpas let their lambs milk the ewes for a month and after that they tied the ewes up in a long line and milked them themselves.

'How do you shear the wool?' I asked him.

'We cut it off with knife,' he replied. This set me wondering what the rate per hundred would be in New Zealand for shearing merino wethers with a knife!

We took him to the woolshed, and a photo shows him watching Ian shearing a sheep with blade shears, despised as primitive in most of New Zealand. Clutched in his arms is little Susan Harrison, for the children were a great delight to him and he to them.

When he left us he took as a present from Grasmere a couple of pairs of blade shears, and these were among the most cherished possessions which he took home with him. Later on I believe Ed Hillary arranged for more of the same kind to be sent to help the Sherpas with their shearing.

The population of the upper Waimakariri Gorge has been small ever since the exodus of the railway construction gangs, and except for the permanent station people it has always been very much a floating population. With the notable exception of Aunty Grace, the well known postmistress at Cass, and her husband, Sam Robertshaw, few of the railway employees stayed very long. Some of them accepted the isolation and hardship of their lives—and to pass a winter in one of the railway houses often was a hardship—because they could get no other job, and a few because they were unpopular with the police and were less likely to suffer harassment where the boys in blue were scarce.

One that I remember had a car—a pretty battered and probably unroadworthy specimen of its kind. Unfortunately hhis numerous offences with it had caused him to be prohibited from driving, which was a restriction not to be borne, and one which was often disregarded. The police from Darfield and Otira knew that he often drove it to the Bealey Hotel, and tried several times to catch him; but he seemed to have a sixth sense which warned him when the enemy was approaching, and he always managed to change places with his wife or someone else before they arrived. One evening he spent a considerable time at the pub, and perhaps his driving was more than usually erratic when he left for home; or else it may have merely been that some weary portion of his ancient vehicle gave way and precipitated him to disaster. Whatever the cause, when he was coming down a steep little hill the car veered wildly to the left and tore through the fence that divided the road from the railway. Unfortunately, the road was some 6 metres above the line at this point and the car dropped with an appalling crash on to the track. Some time later, the early morning railcar, which carries the *Press* and a few sleepy passengers to Greymouth, came sweeping round a bend and ran straight into the car. The driver had no chance to stop and the shock of the collision flung the battered remnants of the car over the other side of the track and down a bank into the river bed. The poor driver was horrified—for all he knew there might have been a badly injured man in it. It seemed unlikely that anyone could have simply walked away from such a crash. He got out and with the help of guard and passengers he searched among the willow trees which lined the track and far and wide in the river bed for any sign of a body. They found nothing, and after delaying for an hour or more the railcar had to go on.

The Darfield policeman had been summoned, and came up prepared to remove a corpse if one could be found. When he saw the wreckage he took a brief look and laughed. 'That's Barrett's car,' he said. 'He won't be dead. Has anybody looked for him in bed?' Nobody had thought of such a thing, but sure enough that's where he was. When interviewed he claimed that his wife had been with him at the pub and that she was at the wheel of the car when it went over the bank, but there were witnesses both at the pub and in the railway settlement who knew—and said—that this was not the case. They had him at last, and on this occasion there was no option but the gaol. But then the age-old problem arose—the public paid the bill to keep him there and to keep his wife and family at Cass as well, and the Railways were a man short in the gang and couldn't replace him because they had to let his wife remain in the house until he returned.

The epilogue was equally ironical. When he was released he drove up in a borrowed car to take his family away, despite the further disqualification he had incurred—and I dare say the police took care to look the other way!

Chapter 20

NEW IDEAS

> Wool falls downward from the ewes as softly
> as snow drifts down the mountain, heaping high
> along the shearing boards as in the valleys,
> drift upon drift upcoiling foot and thigh.
> ARNOLD CORK, *The Golden Fleece*

DURING the 1950s a few bold farmers in the South Island started shearing their sheep in early spring, before lambing. It was not quite a revolutionary idea; I knew a man who did it many years before—in the late 1920s, in fact.

Trevor Barker had a farm in Beautiful Valley near Geraldine and because he had a lot of manuka scrub on the place and burnt it in the spring to suppress it his sheep came in for shearing with black wool instead of white. He decided that if he shore them before the burning season he, like others, could sell nice white wool. There was another reason which he did not advertise so much; he liked to spend a week in Christchurch for the Grand National meeting in August. Once the wool was off his sheep, a week away would not matter much and he could lamb a little later if he had no shearing to do in November. As far as I know he was a pioneer of pre-lamb shearing, but nobody thought his experiment worth imitating in those days.

When early spring shearing did begin to catch on there were some violent protests. It was cruel to remove the wool from sheep in winter or early spring; they would suffer severely from the cold, and horrible prophecies were made about the shocking losses which would ensue. Even Oliver Duff, who wrote for many years in the *Listener* about country matters, berated the sinful greed of farmers who so cruelly misused their stock.

Strange to say, few instances of death from cold occurred and the experimentalists began to find that they reaped a benefit in unexpected ways. The wool—though rather light in weight—was much cleaner and brighter. The ewe, once shorn, went quickly to shelter when a storm arose, taking her vulnerable new-born lamb with her, which nature strangely had never taught her to do, and it was much easier for that clumsy little new-born lamb to find its milk supply when it was not hidden under a canopy of wool. The ewes did not get cast before or during lambing, and so shepherding was not required after the lambs were born. Shearing took place at a time when there was little other urgent work; it did not clash with haymaking or harvest or the sowing of crops. From the shearer's point of view the practice lengthened his working season and avoided the awful rush of summer shearing. Fewer men could now shear the same number of sheep. Surely a list of benefits not lightly to be despised.

Why did the sheep not feel the cold? Here was a question upon which hung the whole argument about cruelty. Some farmers even said that they did not rush for shelter, and went on grazing in the open despite the wet and cold. A study of temperatures at different times of year perhaps provided the answers. In late winter and early spring it was always cold, the temperature variation being only a relatively few degrees; but in high summer it might be 30°C one day and drop, when a storm came up, to 5°C or less the next. This contrast was what the animals could not stand, so losses after shearing in mid summer were not so uncommon as the critics believed. Later on, some tests were undertaken at Lincoln College which indicated that nature, that crafty protector of her creations, actually provided a thickening of the skin in winter, which helped to insulate sheep from the cold. One of the most important factors was of course the method of shearing. The practice led to an upsurge in the old craft of blade shearing, just at a time when it seemed doomed to disappear altogether.

The wetter parts of the high country had always stuck

faithfully to 'the blades', because they left that vital centimetre of wool upon the sheep's back to protect it from the cold. Dry cold does little harm, but an icy wind driving cold rain or very wet snow is sheer murder to a freshly shorn sheep. Not long after the war, experimenters designed a 'snow comb' to try to get the same results by machine shearing. It had limited success because it was hard to drive through the dense, fine wool of high-country sheep, but it was used to some extent in pre-lamb shearing. Some people—including Godfrey Bowen, the Wool Board's chief instructor—saw it as the death knell of the old blade shearing gangs, but in Canterbury at least they refused to die, and pre-lamb shearing did much to preserve them. Even farmers right down in the plains turned to the blades again when they adopted the new system, and when the shearing competitions at the shows begain to include blade shearing in their displays, the skill and speed of the good blade shearer astonished a great many people who saw the old 'tongs' as a relic of the past and nothing more than a good joke in the modern world. Machines were put in at Grasmere by Sealy Rutherford in the early years of the twentieth century, as they were in a number of stations, but they were not used for very long for shearing though they still remained, driven by the big overshot waterwheel, for crutching, when I took over the management in 1930. The same old gear is still in use in 1978, now driven by an electric motor, but I have never been tempted to go back to machine shearing, even with a snow-comb—or what is now called a 'cover-comb'.

The idea of pre-lamb shearing attracted me at once. Shearing in mid summer was always a nightmare, with haymaking and turnip sowing going on at the same time. Before the war men were glad of work, holidays were few and far between, and it was possible to shear at Christmas time. Often we stopped shearing only on Christmas Day and worked on all the other days of the now sacrosanct Christmas period. After the war there was no more of that, and so shearing had to start uncomfortably early, too soon after lambing; or wait till January when we were wanting to get our sheep out on to their summer country on the tops.

In 1953 Mary and I managed to steal five months from the station to make a long-dreamed-of visit to England. That has no part in this story of the high country, except that on our return we were greeted by a call on the loudspeaker in the fine new Christchurch airport, asking me to go to the telephone.

The caller proved to be Sandy Garvie, our shearing contractor for some years, and his message was urgent and alarming. 'Get your boots on,' he said cheerfully, 'we've nearly finished at Hanmer and we'll be with you in a week!'

Down to earth with a sickening thud—we rushed home as fast as we could to get the tailing done before the shearers arrived, and it was this kind of crisis which made me all too ready to try the early shearing. We did it bit by bit—hoggets the first year, two-tooth ewes the next, and so on, till by 1959 the whole flock was early-shorn.

We had some awful, frightening storms in those first few years, turning sheep out on to blocks where there was heavy scrub, and creeping out in the first light of morning expecting to see white corpses everywhere; but to our relief there were very few. Shivering sheep cowered under the bushes, it was true, but they stayed alive—all but a handful. Gradually our confidence rose as the years went by and we had no substantial loss.

The benefits were many, and particularly so with us. Our nearness to the main divide means heavy and frequent rain, particularly in the spring, and once the temperature starts to rise deterioration begins in the wet wool.

In summer shearing we used to get a very high proportion of fleeces stained by what is called dry-yolk—the yellow, egg-yolk colour which is usually accompanied by matting of the fleece. It can be so bad that it becomes almost impossible to tear the fleece apart and it can be thrown on the floor and used like a sheepskin rug. We often had twenty or thirty bales of this cotted wool out of the 200 which we used to shear. All this disappeared with the spring shearing and there were many advantages besides—not least that money came in much earlier in the season. Before that we had to wait till March to see a penny of income for our work, and the whole year's earnings came in three months—March, April, May. It was a long time to wait until the following March to know what your income was likely to be next year.

Another innovation which began about this time at Grasmere was the introduction of Corriedale rams to what had been for at least fifty years a 'halfbred' flock. For the uninitiated, the cross between a merino and a longwool sheep, usually Lincoln, Leicester or Romney, is called a halfbred and it was an article of firm belief among high-country men that using a 'first cross' ram of this unstable breeding ensured a dash of hybrid vigour, as stock-breeders say, to provide the hardy constitution needed in mountain flocks.

The Corriedale sheep was in fact only an inter-breeding of these halfbred strains to produce a sheep which would breed true to type, because any student of genetics can tell you that only fifty per cent of the property of a first cross between two breeds will truly reflect a half inheritance of each parent—the rest will throw to one or other in varying degrees.

The trouble began because the Corriedale was too successful. On top-class country it out-produced the old halfbred sheep and soon became a breed of world importance, in much demand in South America. Naturally the breeders of this gold mine ran it for all it was worth, increasing its size and the amount of wool it could produce under good conditions until they bred it to a state when it could only thrive under those optimum conditions. A few high-country farmers tried it and condemned it out of hand because it could not stand the harsh climate and low quality feed they offered it; so they went back to the well tried halfbred—all but a very few.

Among these very few was Hugh Ensor of Double Hill Station in the Rakaia Gorge which bounds the Waimakariri on the south. He, and his sons in turn, stuck to the Corriedales he bred at Rakahuri in North Canterbury and ended by confounding all the pessimists who said the Corriedale was no good in the high country. Certainly Double Hill is some of the best high country in Canterbury, but I think a good deal of their success was due to the fact that the origin of their flock was an English Leicester cross and not the more usual Lincoln.

The latter cross is perhaps capable of a greater size and maximum production but lacks the hardiness which the Leicester imparts. When we began to use Corriedale rams it was to Duncan Ensor that I went for the most suitable sheep. The results were encouraging at once; a far more even line of lambs made culling much easier and the slightly less leggy animals proved to be heavier than they looked. There are always a lot of late-born lambs and ones whose mothers have been short of milk, which need to be culled on account of their small size. If you add to these a number of larger ones whose wool is much finer or coarser than you wish to keep, culling will take a formidable proportion of your total lamb production. The comparative evenness of the Corriedale wool enabled us to keep all but a few of the larger lambs and sell a bigger number of the smaller ones. Gradually the number of Corriedale rams increased and the older halfbred ewes passed out of the flock, until we were able to call the whole a flock of Corriedales.

We do not find them lacking in hardiness and constitution, though admittedly they are better fed and looked after than were our sheep before the war—sheep are worth much more now, and topdressed blocks enable young sheep to mature earlier. Nor are they less popular with the farmers to whom we sell our surplus ewes; quite the reverse in fact.

In 1965 a long and fruitful association came to an end. The partner with whom I bought the station in 1930, Leslie Orbell, had died in 1955 and I had carried on the partnership with his estate for ten years. Now I hoped to provide for a son of my own to carry on for another generation. It had been a happy association and nobody ever had a better partner than Leslie. When things went well he was delighted and appreciative; when they went ill, even during the big slump of the early 1930s, he never complained or criticised. My one regret was that he did not live to see the success of our topdressing after the fire; that would have given him great pleasure. At least he got the benefit of the high wool prices we enjoyed after the war, a reward he well deserved for his courage in risking his money in a high-country station with an untried manager.

Changing partners was not a simple matter, and it was complicated by a dangerous threat to the whole economy of the place. The leases of some of our high summer country at Cora Lynn were running out and pressure from the North Canterbury Catchment Board decided our landlords, the Lands and Survey Department and the Arthurs Pass National Park Board, not to renew the leases for a further term. We had already lost some land on the Burnt Face, which the New Zealand Forest Service resumed because it was eroded, and this last blow had threatened our whole system of management. After these leases expired we should have no summer country at all—no place which we could use to lighten the burden on the heavily stocked winter blocks. It might mean a drastic reduction in the number of sheep we could carry, with very little reduction in the cost of running the remainder; for the rents were very low and the cost of mustering small on a property where there were musterers available in any case. What was a week's extra work for them? Anyway they all loved the fine, open basins where a man and his team of dogs are pitted against the rugged rocks and streams, the violence of the sudden storms, and the cunning of the free running wethers, always anxious to escape. It was a job every man looked forward to. Those who knew it well enjoyed following the beats which they had learnt to know, remembering the difficult corners where sheep had dodged them in past years, looking for short cuts where they could save themselves a hard climb, or their heading dog a long and difficult run. Those to whom the country was new approached it with the fascinated curiosity which high-country musterers feel towards a piece of country they have heard about, for the men in mustering gangs discuss these things as mountaineers discuss the peaks that they have climbed. There is a freemasonry in this little world of men whose passwords are the vocabulary of horse and dog and sheep and whose mental landscape is the open mountain side, where sheep have to be sought under the shadow of the minor peaks of the Southern Alps. Who among them has not heard of the 'Valley Country' of Mesopotamia, the 'Moa Basins' at Mount Algidus, or our own 'Powers Country' at Cora Lynn? And who among them does not jump at the chance to take a beat on such a well known block and test his skill and endurance and his team of dogs against its unknown challenge? The closing of this lovely piece of country will be a blow to many men besides ourselves; men whose names have been recorded—not in gold letters on some roll of endeavour—but in a few initials on the wall of a mountain hut and in the undying tales which are told round the burning logs of a camp-fire after the day's muster is done. In more than 100 years of use, how many nimble little hoofs have traversed the knife-edged ridge of the Hut Spur? How many hobnailed boots have left their print behind the sheep as the long string wound its way down to the safety of the winter country before snowstorms buried summer grazing under a protective blanket? Many whose names have become well known among high-country men have left their bootprints on those rocky ridges, men like Harold White and Peter Newton, Reg Ferguson and his brothers Jim and Brig, and names from long ago, like Charlie Cran and Fred and Arthur Booth—and who was Power who gave the block its name? I have never found out. It is invidious to pick out individuals. No man who filled a beat on that country has need to feel a slight; each and every one of them well deserved to 'walk by right on the naked hills'.

Is the closing of this country necessary? There I can only admit that I am prejudiced. Except for a few, steep, sunny faces there has been no visible deterioration in my time, and what there has been I believe occurred when the country was overstocked when I first went there. Lightly stocked, and grazed for no more than three months, I do not believe it suffered at all. Far more damage was done in the bush during the heavy infestation of deer which existed up till the 1950s than has ever been done by sheep above the bushline.

However, such an argument is fruitless. Land is now classified according to a code beyond our control and although the administrators of that code are tied by their responsibility to apply it, they are not without sympathy or understanding and I have good cause to give them praise where it is due. They could have said that the leases had expired, and we would have had no legal right to complain, for in the main the land was covered by a Pastoral Occupation Licence which, even under the enlightened provisions of the 1948 Land Act, gives no right of renewal. Instead they looked, as we had to do, at the economy of the station as a whole, and decided to grant us assistance in developing off-site grazing as provided under the soil conservation regulations. Provision of off-site grazing means the improvement of areas not subject to erosion hazards to carry stock displaced when vulnerable areas are retired. It is an intelligent and reasonably generous attempt to solve a very difficult problem, and the Soil Conservation Run Plan under which it is arranged is a partnership between the farmer and the community at large to share the cost of a new system of using the delicately balanced vegetation of the mountains for pastoral production. The farmer provides rather more than half the money required, and suffers most of the headaches and all the anxiety, while the rest of the community provides some cash it never knew it had, and leaves its share of headaches to the professional soil conservators who have little but hope and good intentions to guide their steps in this very experimental field.

The Chief Soil Conservator in our case was our old friend Doug Dick, who had helped us so much after the Bailey fire of 1956, and he and I set out to draft a run plan which would provide for the retirement of the Cora Lynn Tops, the Burnt Face, Bealey Spur and Powers Country, a formidable total of 6000 hectares—give or take a thousand or so, for surveying in the high country was never a very exact science. This meant the displacement of about 5000 sheep, but since they were only there for three months out of twelve the total could be divided by four on an annual basis. Feed for an extra 1200 sheep on 12,000 hectares doesn't sound too difficult to achieve, but the snag is that it is much more expensive than the free open grazing that the mountain tops provide. There is no point in going into the details of our plan; all that needs to be said is that in spite of our combined knowledge and experience the result which has emerged is quite different from the plan we drew, and twelve years later Powers Country had still not been completely retired.

So much for the schemes of mice and men. The trouble is that the system works quite well where the area to be retired is small in relation to the rest of the property or where the rest of the property is a sound economic unit without the retired land; but when the whole management of the place has depended on the use of a large area of summer country—which amounted in our case to a third of the total land—the whole balance of the station is changed and a completely new management pattern must be developed. The pastoral system of the high country evolved by trial and error and is hedged about by the unyielding parameters of the mountains and their climate. You cannot ride roughshod over nature in any type of farming, and in the high country least of all. It was against the background of this plan that we had to negotiate a change of ownership and it was no easy matter to decide what the future profitability of the place might be, and fix a price we could afford to pay.

One of the first things that had to be done after the change was to provide a house for Ian to live in because Mary and I had no immediate plans for retirement and the rambly old house which the Hawdons began in 1858 was so much home to us that we couldn't think of leaving it. As with most of the early stations a series of blocks of freehold land had been bought from the provincial government, surrounding the first homestead and scattered here and there about the most fertile areas of flat land, to protect owners from eager land-seekers before there was any real security of tenure. This was a system known as 'gridironing' because there were often spaces left between the blocks to avoid having to buy too much. The Grasmere freehold covered most of the flat land in the vicinity of the homestead—by no means the best spot on which to have built a house. It did not run up on to the slopes of the hill at all, and Ian decided that he would like to get more sun and less frost, not to mention a better view; he chose the low terrace east of the main house, which had all these advantages and better soil to boot. Unfortunately it was just off the freehold land and we were not willing to build on the university lease. However, there was a little piece of freehold on the flat near the railway station at Cass, which had once contained the long defunct Cass Hotel, the Road Board office—known as the library—and Cassidy's stables in the coaching days.

How Grasmere came to own this land I don't know, and from a pastoral point of view its value was nil. In exchange for this worthless asset the very helpful Board of Governors of the university agreed to let us have an equivalent area adjoining our other freehold and comprising the land Ian wanted for his house. So it was done, and in due course the cottage was built, with a panorama of the whole Cass Valley spread out beneath it and sun beating straight on to it from early dawn until the great dark bulk of Misery cut it off in the late afternoon. Nothing could move in the whole wide basin; neither trains on the line nor traffic on the road nor station work in the paddocks; not even ducks upon the lake without being seen by a watcher from its windows.

There was one snag of course—the wind. The wind that whirls over the Misery ridge from the west and, confined by the narrow Cass Valley, emerges in a series of violent blasts like some great trumpet player spewing out the whole contents of his lungs. It was over this gulch that a friend of ours was flying, in one of the early planes which linked Melbourne with Christchurch—DC4s which could not fly above 4000 metres. She had stood up to try to take a photo of the homestead to give to us when, caught in a gust of wind, without warning the plane dropped like a stone, leaving her pressed against the ceiling. They said it went down 400 metres. Needless to say we did not get our photo!

To protect the cottage from these blasts we had to grow some trees as quickly as possible. It had never been easy to grow trees at Grasmere, particularly on the shingly flats, and most of what we had were burned in the fire in 1956. Seedlings bought from nurserymen had a poor record of survival and we decided to go to Lake Coleridge where the big plantations put in by H. E. Hart when the power station was built had seeded out into the firebreaks. Fine healthy seedlings of *Pinus laricio*, ponderosa and Douglas fir were to be had for the labour of digging them up with a good spit of fertile soil and, when planted at Grasmere, practically every one grew, a result we never achieved with bought trees. What is more they grew fast and provided shelter far more quickly than we could ever have hoped. The result bore out what we had discovered with our topdressing experiments; that the soils of mountain sides are far more fertile than those of the flats, which have been severely leached by centuries of heavy rain. The mountain soils are constantly rejuvenated by the breakdown of rock particles by the forces of nature, releasing the essential minerals for plant growth. Is erosion then not such a terrible scourge as we have been taught to believe? Is it not nature's way of improving valley soils? After all the propaganda which accompanied the passing of the Soil Conservation and River Control Act this is a sobering thought.

Chapter 21

NO BLESSING HERE

> Fire and ice, ice and fire,
> Antitheses of man's desire,
> Tame servants under his control,
> Two demons freedom can unroll.
>
> ANON

THE CONSTANT ANXIETY about fire engendered by our frightening experience in 1956 was not entirely confined to outdoor fires which might destroy vegetation and stock; there was always the danger of house or building fires in an area as isolated as ours.

For a number of years while I was a district fire officer the Forest Service allowed us to keep a high-pressure pump which could be used in an emergency—either outdoor or indoor—in conjunction with a 1800-litre tank on a truck as a mobile water supply. If this outfit could be got at short notice to the seat of a fire it might prevent it getting out of control, but in fact the chance of doing this in time, and often the difficulty of getting it to the site of a fire along the railway, which was where it was most likely to be needed, made it rather a doubtful asset. Other than this there was no fire-fighting unit nearer than Arthurs Pass or Springfield and a building well alight would be destroyed long before any help could arrive. Over the years at Grasmere we had several narrow escapes and in 1961 one really horrible fright.

Mary and I had been to Christchurch and were coming back, accompanied by our daughter Anne, in the late afternoon. Anne was driving and as we came over the Ribbonwood Creek where the whole panorama of the Cass basin suddenly opens to view we were horrified to see a tall column of smoke rising straight up into the air above the homestead. It was an early winter day and the last rays of the sun were coming through the gap in the hills between Mounts Misery and Horrible, throwing the house and buildings into the shadow of the little hill Romulus with its clump of remnant beech trees. The air was dead still, as the chill of a frosty winter night succeeded the dying warmth of the sun and the thin column of smoke rose absolutely straight for 100 metres. Our first thought was that it might be the house and as Anne drove down the slope of that seemingly endless hill we were trying desperately to determine whether the smoke came from it or from one of the buildings behind. A 100-metre span takes in house, stables, cookshop and musterers' hut, and in the shadow of the hill it could be any one of these. To me poor Anne seemed to drive with agonising slowness and when we reached the bottom of the hill there were still 3 kilometres of road and drive before we could hope to know the worst. Minutes were like hours and it was not until we were half way up the drive that we could be sure the fire was well behind the house.

In the end it turned out to be the low, verandahed hut which housed Percy and the tractor driver, which had caught fire from the chimney of the tractor driver's stove stoked to a crimson heat before the lad went to have his tea.

I was sorry to see the old hut go; in spite of its age and the smallness of the rooms the men liked it for its warmth and cosiness. Unfortunately it was such an old building that the insurance on it was negligible and certainly wouldn't go far towards providing another.

The relief was terrific when we were sure it was not the house but I have never forgotten the misery of that five-minute drive with our eyes glued on the column of smoke, and prayers in our hearts for our beloved old house.

We were lucky but some of our neighbours were not.

In 1959 the house at Craigieburn caught fire. Craigieburn—the new homestead built by F. J. Savill about

1910—is more isolated than Grasmere, being down the blind road which ends at the Avoca railway station. We took all hands from Grasmere and went down to see if we could help, but by the time we got there there was nothing anyone could do except roll away a 200-litre drum of kerosene which stood against the storeroom wing at the back of the house. The greatest loss perhaps in that fire were the diaries and station records which had been carefully kept by Walter and John McAlpine since 1917. Well kept records are all too scarce and it is always sad when they are lost.

There was an even more distressing fire in 1969, this time at the original Craigieburn, now known as Flock Hill. Mary and I had been out late, Scottish country-dancing at Darfield. It was 19 July—the very depth of winter—and there was light snow lying on all flat ground and it was freezing hard. We were glad to be home and warm in bed when at one o'clock in the morning there came a loud banging on the front door. While I struggled out of a deep sleep Mary got to the door and opened it, to be greeted by the dishevelled and distraught figure of Doreen Urquhart crying, 'The house is on fire!' Poor Mary of course thought it was our house she was talking about, but it soon became clear that it was their own. She said she thought it was hopeless but had come for help in case there was anything that could be done.

Half way round Lake Pearson we had little doubt that it was indeed hopeless. Flames were leaping high into the air above the trees which screened the house and by the time we got there it was clear that nothing short of a full-scale fire brigade with plenty of water could save any of it. We found poor Gerald Urquhart running around with buckets of water, his huge, half-clothed frame soaked to the skin and with nothing on his feet except socks. Frozen snow melted by the heat of the flames and freezing again into a glassy deathtrap surrounded the blazing building and it was obvious that the most important job was to get dry clothes and some sort of footwear on Gerald or he would get frostbite at the very least. This we did, with some reluctance on his part, for he was suffering from severe shock and stress and probably he could not feel his feet at all.

After a while we found that it might be possible to save the coal shed and the cook's room, which was a wing of the men's kitchen. The house was originally built as men's quarters by Jim Milliken after he got the place in 1917. He had intended to build a house for himself upon the hill behind, but marital troubles and the slump of 1921 prevented that and he converted it to his own residence and let the men occupy the old house which the Campbells had in 1867. The men had their meals in a big kitchen which was part of the new house.

By working strenuously with hoses and buckets we did preserve the cook's room from destruction but it was really more to be doing something active than for any real use the place could be. When the new house was built the whole site would have to be cleared anyway but it was intolerable to stand round in that freezing air and do nothing but watch the flames devour everything. Some furniture and clothes were saved and stacked pathetically in the flickering light of the flames, but in the morning nothing stood erect except a couple of tall chimneys outlined against the winter sky. I looked at them and pictured the same scene at Grasmere, with crumbling stone walls only fit for the brutality of a bulldozer and I felt I could share something of Gerald's agony of mind.

Sometimes fires have their funny side, especially after all the sweat and struggle and trauma are over. There was one several years before the two house fires I have described, which not only had its comic moments but also raised a few odd legal anomalies. This happened while I was the district fire warden; a position which gave me certain powers and authority.

I was out mustering on the Cass Hill when the drama began, as so many others have done, near the Bealey Hotel. After the destruction of the old Bealey in 1937 two reserves—the Police Reserve and the Post and Telegraph Reserve—were made available for subdivision, as I have described in *Kingdom in the Hills*. The man who demanded the subdivision, Fred Cochrane, took no further interest in the matter once we had resolved to rebuild the hotel on its previous site, and the only section which was ever sold was one on a small terrace above the road. This was sold to a dear old ex school teacher, Eileen Fairbairn, and she built a little weekend cottage in the surroundings which she loved so dearly—the wild mountain country on the fringe of the Arthurs Pass National Park. Round her little house she planted native shrubs, and behind it rose the steep toe of the Bealey Spur with its growing canopy of young mountain beech. Nobody could have cherished her tiny estate with more loving care or sought more conscientiously to preserve its natural beauty, so it was a bitter irony that it should be her hand which lit the fatal fire of garden rubbish which so nearly destroyed it for ever. The usual puff of unexpected west wind was probably the cause—it comes so suddenly out of the mouth of the Bealey River opposite the hotel—but in a flash the fire had spread into the broom which competes with the native vegetation all round that rocky promontory.

Somebody rang Grasmere to warn me and of course I was not there. Mary raced over to the Sugar Loaf, knowing that I was likely to be on the bottom beat on its near side, but she couldn't find me anywhere so she raced home again to do my job for me as best she could. It happened that we were shearing, and it happened also that the shepherd from Craigieburn was with us on the hill and he had come up in the Craigieburn truck, which now stood in our stable yard. Our own truck was at Cass, waiting to bring home the musterers after their day on the hill and she thought it better to use the one at hand rather than deprive the men—and the fire warden—of their means of transport by taking their truck away. So, rousing out the shearers, who were enjoying their Sunday rest, she shipped them off in the Craigieburn vehicle to fight the fire, 16 kilometres up the road.

It never occurred to her to examine her legal standing; when there's a fire you go and fight it and leave such irrelevant details to be answered later, and I don't suppose the rights and wrongs of it would ever have been questioned if the manager of Craigieburn had not been another Irishman, Jim O'Donnel.

He, of course, got the message about the fire and like a good high-country man he went to help put it out; but when he got there he was not best pleased to find his station truck standing on the road under no other charge than one of the Grasmere shearers.

Who had organised this arbitrary use of a vehicle under his care? The answer was—Mary McLeod! What right had she to impress his vehicle no matter how good the cause?

Well, what right did she have? She was not the fire warden. Does the fire warden's wife have any right to act in his absence? Even if she has, what are the rights of the fire warden himself? He can impress a man and that man's wife or child—it says so in the Act; what it does *not* say is that he may impress a man's motor vehicle!

Jim took his vehicle back to Grasmere when it was clear that the fire was under control, and he and Mary indulged in a typically Irish argument about who was in the right. The unfortunate shearers who had driven the truck up were left to walk home. I had little to do with the whole affair for it was all over by the time we got off the hill. For that I am profoundly grateful because the last thing anyone wants is bad feelings between neighbours, especially when they have to work together.

There is an epilogue of course—all good stories have an epilogue—and this one was a sad, pathetic little scene. Someone surveying the extent of the damage saw the little figure of Eileen Fairbairn, armed with a silver teapot, pouring water from its spout on the few smouldering embers of her cherished garden, while behind rose the ravaged background of the spur above her cottage, its once lovely, dark-green beeches blackened and still smoking right up to the skyline.

The Forest Service has the power to prosecute those who light fires carelessly. It would have been too cruel if they had done so in this case.

One of the changes brought about by the run plan was in the employment and use of labour. For many years after I went to

Grasmere we followed the traditional pattern of the high-country stations, cutting labour down to a bare minimum in winter and engaging seasonal musterers for the summer. This usually meant that the skeleton staff had their meals in the house in winter, and in the cookshop with a man cook in summer time. But cooks were ever hard to find and, as some of the stories I have told suggest, many of them presented more problems than they solved. With the gradual retirement of the summer country and the more extensive cultivation of the flats for winter feed the balance of work shifted towards the winter, and with the erection of fences to subdivide the blocks mustering could be done with fewer men. The 7-kilometre fence along the top of the Ewe Country meant that each of the two parts could be mustered easily by three men instead of five for the whole.

We already had the cottage built for Ian when he got married in 1965, and if only we had another to house a married shepherd or tractor driver we dreamed that we might be able to dispense for ever with a station cook, except at shearing. What bliss we thought that would be! Building cottages is an expensive business and we didn't feel we could afford to do it, but in 1969 a sudden chance presented itself. The Railways Department was having great difficulty in staffing its wayside stations, particularly the more isolated ones like Avoca, Craigieburn, and Cora Lynn. The old days, when maintenance of the line was carried out by small gangs of surfacemen living in tiny groups 11 or 12 kilometres apart and travelling up and down the track on hand-pulled jiggers, were dead. Motor trolleys made them more mobile and most of the repair work was now done by gangs sent out from Christchurch in a bus; only at a few strategic points like Cass and Arthurs Pass were there any permanent staff. So they began to sell the old railway houses for removal. There were people who would readily have bought them on site for weekend baches—particularly in areas where there was deer-shooting or fishing to be had, but the rule was that the house had to be removed from railway property. The result was that they were going very cheap because of the high cost of moving them and the lack of any place to move them to. Craigieburn bought one of the houses and then were allowed to use it where it was, much to the fury of several competitors who would have raised their bids if they had had the same advantage. There were none for sale at Cass, but there were two at Cora Lynn and both these were available for tender.

Roger James of Benmore Station heard about them first and he and I went to look at them together. I knew they would be in quite good condition because they were built much later than most of the houses up the line. The first ones at Cora Lynn had been built right back against the bush half a kilometre from the siding. The railway runs on the cold side of the river here and in winter these houses only saw the sun for about two hours in the middle of the day. They were not insulated in any way nor was the plumbing designed to withstand frost; in fact it would have been difficult to find colder or more inconvenient dwellings in the whole of New Zealand. Small wonder that the Railways Department found it difficult to get people to live there. In the end they had no option but to build two new cottages beside the line, where at least the men were closer to their work and the sunshine hours were more than doubled.

Roger and I examined these newer cottages carefully, giving our attention mainly to the question of whether they could be moved. I must admit there was a formidable list of obstacles. First, they were far too big to be moved by rail unless they were completely demolished; second, they were on the wrong side of the line and would have to be got across it; third, the track here was well above the river bed, on the toe of a large fan which had been cut into a steep terrace by the floodwaters of the Waimakariri; fourth, the passage across the line was blocked by the same set of multiple power and telephone wires which had obstructed us at our own power house; fifth, there were 3 kilometres of shingle river bed to negotiate before the south bank could be reached, across which the changing streams of the river meandered at their will, sometimes combining into a major river which could be deep and swift; and finally, the road on the south bank was at least 10 metres above the river bed and lined with another set of telephone wires. It would be difficult to find a job which presented more complicated problems.

There was only one thing to do and that was to call in an expert, so we sent for Jimmy Curline whose regular occupation was moving houses. We had imagined that each house could be moved in one piece because they were not very wide, but he soon disposed of that idea, if only because the West Coast Road is not wide enough at the crucial point where we had to get on to it.

No, the houses would have to be cut down the centre line of the roof and reassembled at their new site, each half being carried on one of their big transporters. Except for this he believed the job could be done if—and here was the biggest IF—the weather behaved itself and did not send the Waimakariri down in a raging torrent at the crucial moment.

There is no need to describe all the negotiations that went on. The houses were bought for the proverbial song; a track was bulldozed down the wide river bed; railway staff were booked to supervise the crossing of the track and the lifting of the wires, and post and telegraph linesmen to deal with the telephone lines on the other side.

Early in May 1969 the cavalcade assembled, the houses were sawn in half and jacked up high enough to get the big transporters under, and the great manoeuvre was ready to begin. The early winter often provides a period of stable weather and the freezing of the streams which feed the river reduces its flow. While all our preparations went on nothing happened to disturb our tracks or make the main stream impassable. When the great day dawned there was no rain to interfere with us but a dark cold mist hung over the whole valley out of which occasional snow showers drifted down to whiten the ground.

The big transporters lumbered off, with the bisected houses rocking alarmingly on the rough track. They looked exactly like a child's doll's house, where the front is complete but the rooms are open at the back so that furniture and the doll population can be manipulated by the child; although in this case there was neither furniture nor inhabitants and only doors and windows and cupboards and torn linoleum showed what the rooms were like to live in. Each half house was 12 metres long, 4 metres wide and 6 metres high on the transporters, which gives a fair idea of the difficulty of manoeuvring them.

The first obstacle was a deep ditch beside the track which had to be filled with old sleepers, of which there were fortunately plenty erected as a wind break on the north-west side—the houses must have been shockingly exposed in any case. Once over this hollow and the track itself, there was the line of poles with the power cut off and the wires raised so that the roof passed safely under; then the steep drop bulldozed down the terrace, with the river bed itself complicated by springs which come out from under the fan above; then the long haul down the whitened river bed, twisting and turning as the smaller streams were crossed and then, late in the afternoon, the crossing of the main stream where it tends to run hard against the southern bank.

We had been very anxious to begin with, but the skill and experience of Jimmy Curline and his crew bred increasing confidence as we watched the sections, which looked so huge as they were dragged across the line, dwarfed to pygmy size by the great expanse of the river bed and the towering mountains which surround it. The transporters churned through the river one by one and then there was only the last obstacle—the ramp which the bulldozer had had to build up on to the West Coast Road at the foot of the hill known as Paddy's Bend. At the top of this ramp were the multiple wires of the main telephone line, with the patient linesmen standing by to raise them above the roofs. Up the houses came, twisting at the top perilously close to the sheer cliffs of broken rock out of which the road is cut. There was only one more anxious moment, when the cavalcade had to pass what is known as Lookout Point, where there is very little space between the clean-cut rock face and a drop of 6 metres on to the railway track. It was here that Barrett's car had gone over. In the coaching days this place was known as 'Cornishman's Rise', though who or what the Cornishman was is lost in the irrecoverable past. Paddy's

Bend is supposed to mark the place where a horse fell over, so perhaps the Cornishman suffered the same melancholy fate. In any case the Cora Lynn houses survived this Scylla and Charybdis and proceeded up the more open hill over Goldney's Saddle. That was enough for one day's work and they were left beside the Cass Creek to wait for another day.

The next day we got the cavalcade to its site at the bottom of the Bailey terrace after knocking out a couple of gateposts and cutting three fences. Only then was I told that they needed a concrete mixer, a metre and a half of shingle and the requisite quantity of cement *at once*! Of course our concrete mixer engine would not go, nor was there the proper belt for it and we had no cement. However, after hasty repairs the engine functioned; a modified platform enabled us to use a belt off the governor at the power house; I 'borrowed' two bags of cement from some M.O.W. builders at Cass and a convenient stockpile at the Cass Creek—also M.O.W. supplied the shingle. We ran round for a couple of hours organising all this and cursing Jimmy Curline for not telling us before that he wanted it. The men soon made up for lost time and worked away like beavers, putting in piles till long after dark.

The third day saw the two halves of the house lowered on to their piles and the astonishing feat of joining them together again so that it was almost impossible to see that they had ever been divided. Of course this was far from the end of the operation; painting, plumbing, papering, drainage and cooking arrangements all had to be completed, and the old washhouse which had been separated from the house by a wind tunnel was made to join on beside the back entrance.

When it was all finished some months later it made a nice cottage but the original price of a 'song' had swelled to a very sizeable total. Nonetheless it was a fascinating exercise and we certainly couldn't have built a new cottage for the same money.

Chapter 22

IRRIGATION

> Then shall the lame man leap as an hart, and the tongue of the dumb sing: for in the wilderness shall waters break out, and streams in the desert.
> *Isaiah*, 35, 6

WHEN R. D. Dick and I designed the soil conservation run plan for Grasmere we failed perhaps to appreciate fully the effect of a major change of system. The limitation imposed on stock numbers by a conservation-conscious administration was rightly based, under the old extensive grazing system, on the winter carrying capacity of the land. The level of winter snow and the long months of icy cold in which the grasses do not grow had always determined how many sheep a run could carry. If the grasses grew consistently in summer time there would be no problem then, but in actual fact they don't. In perhaps three years out of five there can be a longer or shorter period in summer when the flats and lower hills are parched with sun and searing nor'-west winds and the growth stops as suddenly and as completely as it does in winter. It was because of this that high-country stations were driven to use the inaccessible mountain basins where cloud condensation maintained the moisture level even in mid summer. Take those basins away and the summer drought can starve your stock as inexorably as any snowstorm.

What our plan had done was to transfer the period of danger and anxiety from the winter to the summer; and in fact it compounded any winter problem that there was, because it envisaged more cultivation for hay and turnips for winter feed and these crops in turn were threatened by a summer drought. If it went on too long they would not grow at all!

Not surprisingly, perhaps, old Mother Nature chose this moment to teach these presumptuous mortals a lesson and she dealt out a series of summer droughts which forced us to scratch our heads for new ideas and keep on using Powers Country to bridge the gap. The head-scratching produced one new idea and at the same time the significance of our problem, and that of other large high-country properties which had adopted run plans, began to be appreciated by the catchment boards concerned. They came to realise that the changes in management which they desired to see were costly and often unsuccessful unless the assistance was much more generous than they had thought necessary. There was a grave danger that unsuccessful run plans would give the whole project a bad name among runholders and result in their complete refusal to participate. The scheme which emerged from our head-scratching exercise was an ambitious one which had the advantage of attacking the problem of summer drought directly. It was to use the Cass River behind the Grasmere homestead for a private irrigation scheme. We had always had the 'water race that runs uphill' to provide stock water in the paddocks and from time to time this had been used to flood small areas of ground to stimulate the grass. We knew the potential was there all right but what were the economics of such a major job? My son, Ian, who had by this time taken over the general management of the place, was the initiator of this project and upon his head descended all the problems which arose in its construction and the adjustments which resulted from the constantly rising cost.

Luckily at just this moment the Soil Conservation Council changed the basis of its grants. Instead of subsidising the development of off-site grazing to substitute one stock unit for each one displaced by retirement, they agreed to give grants for twice the number—so helping to compensate for the extra cost of this sort of management. This meant a reassessment of the whole of the benefit to which we were entitled under the run plan and our attempt to estimate the extra number of stock which irrigating would permit. More frantic guesswork was required for there was little practical evidence to base such calculations on.

The ultimate conclusion was that we could irrigate 180 hectares, paying rather more than half the cost ourselves, at a total cost of between $30,000 and $40,000.

A survey indicated that there were over 400 hectares which could be watered by border dyke and, since the land fell all the way from the homestead to the lake and out towards the Cass River on the north, we had a considerable area from which to choose our 180 hectares. However, that choice was a long way ahead; the first hurdle was to get a water right, and that's an exercise well hedged about with difficulties in these enlightened times.

Long years ago I had a dispute with the Ministry of Works over the culvert which carried the water race across the main West Coast Road. It got blocked, and flooded the road, and they wrote indignantly to say that I did not appear to have legal easement for the culvert. I don't suppose there ever was one! In the happy days long ago the race no doubt ran in a little channel across the shingle highway into which the coach wheels dropped with a sickening thud and the driver cursed just one more of the natural obstacles which beset his path between east coast and west. Then one of the many roadmen who patrolled the road with horse and gig and wheelbarrow and shovel put a few pipes in to pacify the early motorists whose vehicles were less rugged than the horse-drawn ones, and so the culvert came into being—no names, no pack drill, as the soldiers say. If you don't ask permission, is not refused! On this occasion I disregarded the official letter and arranged the matter amicably with the road foreman; the culvert was fixed and nobody even noticed that the letter went unanswered. Nowadays it's not so easy. Protesters are the order of the day, and anyone who wants to do anything must run the gauntlet of their scrutiny.

We had little fear that people like the Ministry of Works would try to spike our guns provided that we complied with their requirements about crossing roads, but the Acclimatisation Society was another matter. We had already had a brush with an overzealous ranger over several small fish which died in our power house race when the change was made to the new pipes, in spite of having rescued as many as we could from the muddy weeds at the bottom of the old channel; so we had little doubt that suspicious eyes would be on every move we made to divert another stream.

The Cass is a swift little mountain river and ever since the fires which raked the forest from its steep sides in the last 100 years it has scoured its rocky bed with every passing flood and built up a wide bed of loose and permeable shingle when it reaches the more open ground. Here every summer when the flow is low and evaporation high the stream, dignified by the name of river, disappears underground and doesn't reappear till after it reaches the bridge over the main road and is again constrained between two hills. A few salmon spawn in the loose shingle beds where the Grasmere Creek comes out, but higher up, above where the stream bed dries, fish do not live; so it was difficult to see what harm our project could do to even the most fanatical fisherman.

Well, we got our permit with no more restrictions on it than we had expected and the next business was to design the scheme. A costly survey had to be made and then the decisions taken as to where the races should run and what paddocks should be irrigated. The best land lies between the main road and the lake and its outlet stream—the power house supply—so it was decided to do that part first; but when the plans were drawn we found an unexpected snag; all the fences were in the wrong place and running at wrong angles and almost every one must be replaced. This was a cost we had not bargained for and one which was not covered by soil conservation subsidies, so from the very start the estimated cost began to climb. When the work actually began and the giant machines began to level and make the borders, other difficulties gradually appeared. First, the seemingly even slope of the fan was in reality a series of rolls across the slope. As erosion in bygone centuries had carried the shingle from the Cass River down towards the lake in successive waves, these waves had varied in intensity, so that a wide strip or bar of more recent date covered the layers below and formed a hump. Between these humps lay the best and deepest soil but when the borders all ran towards the lake some were much higher than the ones alongside, which necessitated the moving sideways of a lot of topsoil and the baring of rough shingle underneath.

Another problem which faced the draughtsmen was an obvious one resulting from the fan being narrow at the top, and widening at its base beside the lake to a distance of two kilometres or more. So there was a constant need to increase the number of borders, and awkward little triangles appeared which needed special design and treatment.

How far would the water flow? It was necessary to guess at this so as to determine the distance between head races. If we set them too wide apart the water would not reach the bottom. There were few precedents in land like this, because, although the slope was greater than is common in irrigated lands, the soil was very porous in places. Each head race had to be fenced on both sides so that there could be controlled sheep-grazing along it, with no access for destructive cattle, and the fencing bill climbed alarmingly in consequence.

Then there was the design of the intake to be considered. The Cass is often a wild and stormy valley and, although much of it is still full of stabilising bush, the lower valley immediately above the homestead was stripped of its cover by the early settlers of the country—and most probably by the Maori colonists before them. The cover has improved a great deal in the fifty years since I saw it first and I believe the river bed is much more stable than when I went to Grasmere; but a flood can still tear the banks and shift the stream from side to side, depositing great banks of shingle in the mouth of a race and even far down its course if it is not shut off. The new race had to be sited further up the stream than the old stock-water race because, on account of its size, it had to pass the other side of the homestead buildings. At the same time the stock race had to be preserved to supply water to paddocks which were not being irrigated. It is better to have an intake which is not too massive and to draw the necessary amount of water to it by simple channels in the river bed rather than have the main river with its destructive power running near the concrete structure.

When the main feeder race was made there was the usual chorus of disbelief, so deceptive are the levels of these streams. Take any stranger to the spot and he will swear he is walking downhill as he goes towards the intake. In fact, like our friend the 'race that runs uphill', the water flows quite swiftly in the opposite direction.

It would be tedious to describe the many aspects of this project. Contractors came and went; surveys were made and found wanting; concrete sills and dams were built galore—and all the time someone must organise the work and someone must cook and cater for the unpredictable numbers of extra men. Our sons do not enjoy the quiet simplicity of the life we lived in the years of long ago. Their rewards may be greater at times but so is their effort.

There was one more major problem to face. The designers of the scheme believed that if the main feeder race was wide enough it would not scour, even where it ran straight down the slope, as it had to do to reach the road and cross into the lower paddocks. Up at the top near the homestead the rocks and shingle which underlie the fan are fairly large, and even where the race is swift they do not scour, but further down where the soil is fertile and there is little shingle the scouring became serious as soon as a substantial flow of water was turned on. This was an awkward problem and could well prove very expensive. The normal cure for such conditions is a series of dams; but the cost of that was rather frightening, so we had to compromise by filling the bottom of the race for 1 kilometre or more with large boulders.

Fortunately the Ministry of Works had a stock of them where they had been screening shingle at the Craigieburn and Ribbonwood Creeks so at least there were enough available. The land preparation had to be done in stages, 40 or so hectares at a time, interspersed with periods when we did not have the money to continue. Even now (1978) there are still 60 hectares left to do but what we have done has gone a long way towards securing our summer and our winter feed. From now on this project *must* succeed because the lease of Powers Country has expired at last and left us with no alternative grazing.

There was a grand official opening attended by members of the Catchment Board and those who had been associated with the scheme, and a few of our neighbours came along to look. It was a lovely, sunny day and the visitors were able to enjoy having their pictures taken against the background of the intake headworks with the blue water of the Cass pouring through it and the dark-green bush slowly creeping to reclothe Misery whose jagged ridge forms such a lovely backdrop to the station buildings.

As usual at opening ceremonies the subject had been in operation for a considerable time and many of the problems had been met and mastered, though some still remained and still do. Also—as usual—the opening called for the consumption of some very different liquid from that which passes through to the intake grate and makes swift passage down to the road, where it plunges into the large concrete pipe which carries it under the seal, unnoticed by the speeding cars. This kind of consumption often leads to a little high-spirited exuberance and today was no exception. Somebody bet or challenged one of the local shepherds, Tony White, that he would not follow the course of the water and dive through the pipe under the road. Certainly it was summer and the prospect of a cold bath was no deterrent, but to plunge head first into that pitch dark pipe must have taken all the Dutch courage in the world. He did it—and probably emerged cooler in the head than when he went in, but nobody offered to imitate him!

It may seem strange that those who have dedicated their lives to walking on the naked hills should accept the trammels of an irrigated farm, but these changes must needs be accepted if we are to preserve the things of greater value: the right to farm the hills that surround this little patch of exotic cultivation. I am sure they look down upon it all with lordly tolerance, knowing they will remain lofty and aloof no matter what man does scrabbling in the dust at their feet, and secure in the knowledge that the little green oasis could not survive without them; if for no other reason than that they provide the water without which it could not live.

Chapter 23

FAREWELL

> Across the margent of the world I fled,
> And troubled the gold gateways of the stars,
> Fretted to dulcet jars
> And silvern clatter the pale ports o' the moon.
> FRANCIS THOMPSON, The Hound of Heaven

FRANCIS THOMPSON, who wrote these beautiful and prophetic lines, was a man to whom Christ appeared in very truth to be a Saviour, one whose loving footsteps dogged his own and those of all mankind. Men who live among the mountains are constantly reminded that nature is always more powerful than themselves; a knowledge which engenders among them a feeling described by Arnold Lunn—one of the pioneers of European skiing—as the 'religiosity of the mountains'. I believe that human beings have always invented for themselves religions which were within their capacity to understand and from the earliest times they seated their gods on mountain tops from which they could look down upon the puny mortals below, and from which emanated the devastating forces of nature which governed, and still govern, the lives of men. Perhaps this explains why new, psychological religions are arising today, in which the expanding mind of modern man can wander, untrammelled by actual events like births and deaths and crucifixions, among the expanding horizons revealed by the discoveries of space scientists.

Man's need for a religion stems from his knowledge that he is vulnerable to something larger than himself, to forces against which he cannot defend himself except by going on his knees and begging protection from a greater power. Men who congregate in cities seldom see the forces that threaten them. They see only the works of their own hands and they come to believe that with them they can conquer nature, and this belief is strengthened by their relative success in doing so. Nonetheless, at the back of all human minds is the knowledge that no matter how many facets of nature they appear to conquer there are always more beyond their scope. And so they 'trouble the gold gateways of the stars' with invading rockets, hoping by this means to atone for the fact that they cannot subdue the ramparts close at hand.

Those who live among the mountains have no such illusions; day by day they and the animals that they have tamed do battle against wind and water, fire and flood, heat and bitter cold. What modern scientific aids they now employ are strictly dispensable; without these aids they could still live, they and their animals, in their wild and primitive landscape under the shadow of the mysterious habitations of the gods, as the Sherpas and other mountain tribes have done for centuries; and from this knowledge derives the strength which supports them. They do not seek to conquer nature, only to live with her upon her terms, learning by trial and often costly error what she will tolerate and what resist.

This has been the history of the high-country stations for more than 100 years—the gradual learning of nature's rules in a land where there was no pre-history to guide. Costly mistakes were made in the process; some by the runholders themselves—like overburning and overstocking; others in which they had little hand, like the introduction of rabbit, deer, opossum, thar and chamois. The second century has opened with a wiser appreciation of what is at stake and the means by which that stake may be preserved. Grudgingly at first—as will have been apparent in this story of a typical high-country station—but ever more widely, the self-discipline which is required to make us obey the rules of nature has been accepted as a necessity, and with this discipline and our courage and ingenuity we can look forward to another 100 years in which we have no baser vision than to 'walk by right on the naked hills'.

FAREWELL

We have no need to flee 'across the margent of the world', for the world we have is deeply satisfying. As I look back over my years in the high country a hundred vivid memories crowd my mind, like pictures hanging on a wall, over which one lingers lovingly and passes on. I see a figure on a raw young horse climbing the crest of Porters Pass to enter quite alone into an unknown world. A world in which every sight was new, from the soaring crests of the great ranges on either side to the tiniest plant which grew beside the little-used mountain road. I see the reflections of the curious glaciated hills upside-down in the waters of Lake Pearson, the conical lines repeated in the hourglass shape of the lake itself. I see the suddenly unfolding panorama of the Cass basin with the deep mysterious recesses of the bush-clad Polar Range closing it in behind; and as I rode into that panorama with eyes and ears for every sight and sound no whispering in my ears told me that this land would constitute my whole life.

There are other pictures of those early years: unforgettable pictures of camps and huts set against the background of the dark-green 'birch bush' where musterers and their many-coloured dogs idled in the sunshine after striding the shingle tops from dawn till afternoon; the daunting picture of the beat through 'the rocks' at Mesopotamia, which was my introduction to the rugged country of that romantic place; riding at breakneck speed down a steep and rock-strewn spur to try to stop the wild black cattle, wondering at every stride how long my pony's flimsy legs could stand the sickening drops without breaking; the sight of a small group of sheep standing on the steep flank of Mount Bernard at Craigieburn with the dark shape of my heading dog Laddie behind them and the feeling of pride and affection that was uppermost in my mind, for I had sent him on what I thought was a hopeless chase.

These are old memories whose reality is only in my mind. After I came to Grasmere I often wrote down what I saw and felt and have the record to reinforce the memory. One trip to our Top Hut for the autumn muster of Powers Country is recorded in my diary:

'After an overcast morning with low cloud hanging over the lake we emerged into the Cora Lynn basin with a vision of every ridge and spur clear-cut yet coloured with infinite delicacy, tone contrasting with tone and sunlight with shadow, with no harshness, only a pearly kind of beauty breathtaking to see. High clouds rose from behind the ranges in the form I call "fire behind the mountains", as if only one's own valley was spared and fire consumed the rest of the world, sending its columns of smoke up all round.'

My trips to the Top Hut were almost becoming melancholy now. The track unbearably beautiful and unspoilt; the horses and dogs and the whole procedure of an autumn muster unchanged and full of the deep romance it has always held for me; but each year was bringing me nearer to the last, and each year widened the gap between me and the men who really do the job. They still talk of horse and dog and station and the men of the hills who make up our community but, horror of horrors, they bring transistor radios in their swags and drown with shrieking American cacophony the limpid drops of pure music which the mockies let fall round the hut.

'We mustered the Blind Spur and Jordan Spur. Such a familiar day, with north-west clouds trailing their skirts across the ridges and every now and then coming down in a thick blanket which shrouded spurs and basins and mobs of sheep, leaving you helpless and groping. In the intervals, sunlight lit the smiling basins with their carpet grass and celmisias and the dry, warm scent of musk from *Celmisia glutinosa* filled the air and lungs with pure bliss. Across the valley the sharp sounds of barking dogs and an occasional bleating sheep came clearly in the pauses of the wind. I plodded up the creek, headed some lazily moving sheep in the head of the valley and climbed over the ridge of the Jordan Spur to see the magnificent panorama of the Jordan basins spread out beneath a canopy of fog. Down the Waimak heavy rainstorms seemed to come as far as the Anti-Crow and stop, and so it remained throughout the day.'

In 1954 we had a memorable visit from Dame Flora MacLeod of MacLeod, the chief of our Scottish clan. The New Zealand MacLeods had planned to give her a painting to take home as a memento of her visit and the scene was to be chosen by the chief herself. She had seen something of the South Island before coming to us but had not yet seen the view that quite fulfilled her wishes. I drove her to the eastern end of Lake Grasmere and, standing there, she said without a moment's hesitation, 'That's the view I want for my picture.' How wise a choice she made and how delighted she was when Austin Deans painted it for her! I later described the spot in my diary, as I had seen it during a cattle muster:

'I never saw the "picture" spot at the end of the lake more vivid or more varied. Every shade of delicate colour one can imagine seemed to be somewhere in the reflections in the water and it was not completely still, so that the outlines of these intricate mozaics were faintly blurred. Here and there a bird made a pinpoint of reflected light like a diamond winking, and one flew over the surface where it was dark green, beating the water in a series of accurately spaced flashes of light which looked completely mechanical. On the wire fence running into the lake a row of eleven shags sat in a descending line until the last one was almost in the water. Beneath them but upside-down their counterparts made a drooping line in the opposite direction. They must be the full population because I have seen eleven before. Perhaps twelve is an unlucky number for shags!'

One day I rode up to the Top Flats to bring some sheep down. Snow was lying thinly on the fans and the clouds shut down on the tops with a thin drizzle from the north-west. I rode through a fringe of the bush on the other side of the Jordan and was delighted as ever by the spacious grandeur of the forest where the trees are tall and widely spaced. At this time of the year the floor is almost covered with the most vivid green carpet of moss and the seeping rain made everything dark and whispering with moisture. A tree cracked and fell somewhere in the inner sanctuary of the bush and it was all rather eerie and awesome. On the edge of the bush overhanging the tortured shingle were many broadleaf trees, browsed to the highest point of reach by deer, and I reflected how old these trees must be, as there are no young ones and the deer have been with us for fifty years and numerous for forty. It is not only the handsome broadleaf which has suffered such wholesale destruction; toatoa, the mountain celery pine, has had a fatal attraction for stags. The Maoris knew the value of its bark as a dye and the red deer stags soon learnt to stain their antlers by ripping and thrashing its slender stems to get the rich brown colour from the bark.

One of the delights of mustering in the high country is the ever-changing native vegetation. The beech forest—or birch bush as we mostly call it—is usually composed entirely of mountain beech; only occasionally in the Waimakariri do we come across other species. There are several substantial stands of red beech on the north side of the river but none on our side either at Grasmere or Cora Lynn. Every now and then, however, you strike, in the depths of the forest, an unexpected species. A group of lancewoods lies across the track up which we climbed to muster the Burnt Face and, scattered about the one which leads down from the Jordan Basins to the Top Flats, are a number of tall kaikawaka trees whose conical tops stand up above the lower canopy of mountain beech and whose trunks, with their stripping bark, seem to be spirally twisted like a giant corkscrew.

Further up the river again at the foot of the Anti-Crow Spur there are quite large totara trees hidden among the dense thicket of young beech. In the more open country above the bush, whole fields of *Celmisia*, from the huge *Coriacea* down to the tiny delicate *Gracilenta*, can make one's path among the rocks delightful and interesting instead of trudgery. I remember one walk across the Burnt Face in summer where the whole great slope was studded with clumps of *Gentiana corymbifera*, the white flowers standing foot-high upon their brown stalks and seeming to burst upon the world with the most joyous vigour. I came home entranced and dreaming of the fortune which awaited the nurseryman who could combine this lovely plant with the deep cerulean blue of the Swiss gentian *acaulis*.

Some years suit particular plants, and another time I remember well was a shearing muster at Christmas when everywhere I walked I found a dozen penwiper plants. This lovely little rosette grows in deep, cool shingle where its long roots can wind down into the moisture lurking underneath and feed the huge flower head which rises and scents the air as strongly as carnation or gardenia. They were everywhere on the spur that runs off Little Bailey and in the rough borders of the Ribbonwood Creek. Why did they all appear that year? I never saw so many before or since.

In some years the road to Cora Lynn is lined with the tiny, starry, golden flowers of *Corokia cotoneaster* and in the autumn these same drab little bushes are ablaze with orange scarlet berries. Even the inhospitable matagouri contributes its delight in spring when its insignificant little flowers combine in their thousands to scent the air with a perfume which seems inconsistent with the harshness of its source.

Once when I was toiling up towards a saddle on the Cora Lynn Tops I came on a struggling strip of mountain beech clinging to a narrow spur between two great rivers of shingle. The stunted little trees clung gallantly to their remnant of soil and I paused in their cool shade to wipe the sweat off my streaming face; there, at the foot of a little mossy trunk, I saw two brilliant blue puff-ball growths, their colour stolen from heaven above and glowing like little lamps against their soft green background.

One of my favourite beats was a riding one, of which there are not too many on our steep country. When Goldney's Saddle is mustered—the Goldney brothers were the first owners of Cora Lynn—three men go to the nor'-west end to muster above the main road and river to the Cass Bridge, and one man rides across the Cass towards what we now call the power line gully and musters towards them, covering the steep faces of Mount Horrible and a strange mixture of swamp and scrub-covered spurs in which there are infinite hidey-holes for sheep. My faithful little black horse Ponto knew this beat as well as I and we seldom traversed its wandering paths without seeing his deep-marked hoofprints in the softer ground still showing from the last muster.

The beat begins at a large swamp at the head of which the

Goldneys built their first small hut. It was the nearest point of access for their run, most of which was in the upper Waimak Valley, the other side of Mount Horrible. It was no doubt a pretty, smiling place in summer, with the fertile swamp to feed their household stock and plenty of good wood and water all around, but in winter it is as cold and inhospitable a place as one can find and it is little wonder that they moved as soon as there was a road round the bluffs above the river. Now there is nothing to record their presence except a patch of gorse and the discernible line of a track on the far side of the Cass River.

One of the changes I have noticed in my years at Grasmere is the increase in many kinds of birds. I imagine this is due to the cessation of the regular burning which destroyed so many of their nesting sites and the gradual recovery of the beech bush which was either destroyed or cut back by fire. Once on this Goldney's beat I saw a pair of the yellow, New Zealand bush canaries—a bird I never saw before; and often at one particular place at the head of the long swamp I used to see a pair of terns flying wildly about without planing at all, like huge swallows with their almost spike-like, tapered wings. They looked very excited and I could never make up my mind whether they were feeding or protesting against my intrusion into their domain.

The faces on Mount Horrible were quite open when I went to Grasmere and there was much work to do to get sheep off them, for they are very steep below forbidding black cliffs and it was hard to steer a tired dog up through the patches of dense scrub. Now it is very much grown over with dark, half-blighted manuka out of which tall, silvery clumps of toetoe stand gloriously in the sunlight. On the low spurs where one can still ride easily among the scrub, the young beeches are springing up in ones and twos and little clumps. Another 100 years will see this a new forest, and I say *new* deliberately, for there is no sign here that it was forest when the Goldneys took it up in 1860.

We have not stocked the Cora Lynn Tops for many years now. These tops comprise the Misery ridge and that portion of the Black Range which runs from the Cass Saddle to the Burnt

Face. The Misery ridge is an inspiring place to be on a fine day, lying athwart the main valley of the river with superb views in all directions; but with a cold nor'-wester blowing, as it often does, it is as inhospitable a place as you could find. There is no nook or cranny where you may escape the blast, and of all the blocks we used to have to muster it was the one where waiting was the most common. Five men all went by different routes to come out on the top and converge on the spur down which the sheep all had to go, and it was almost a certainty that one or other would be late and keep his mates in freezing misery. Three of the men went up the Cass Creek, crossing and recrossing as it was necessary, and trying to keep dry-footed on the way. Here many years ago we saw a pair of the beautiful blue duck bouncing down over the rapids in a narrow gorge and squeaking protestingly like an unoiled iron gate. One of our shepherds, Ernie Percy, had a prepared plan for this expedition. He kept a pair of worn-out boots specially for the occasion and walked deliberately through the stream at every crossing, instead of leaping precariously from rock to rock. When we got to the open basins at last and paused to boil the billy and eat a sandwich before separating, Ernie would sit down and remove his sodden boots and, from his lunch bag, produce dry socks and his working pair. Each year the pile of Ernie's boots increased at that little picnic spot and their rotting remains probably still lurk among the thicket of mountain totara.

I usually went up the Long Basin, the central position of the muster, which would bring me out on to the top above the great shingle basins which looked down on Cora Lynn. In the centre of this basin I came one day upon a flock of keas. As near as I could count there were between thirty and forty of them flying round a rocky spur which was sunny on one side but which threw a deep shadow on the other and below. The wheeling scarlet wings shot momentarily into the blinding light, where they glowed like burning brands which were as suddenly extinguished as they plunged into the shadow. The kaleidoscope of red and green turned and twisted, swooped and soared against a background of almost black shingle and the pale green carpet grass which covered the outcropping rock. It was an unforgettable sight and kept me spellbound for so long that I had to go racing for the top up the last steep slopes so as not to keep my companions waiting.

There is one last diary picture, taken from a dog-trial day, and then I have done with memories; but I never cease to rejoice that fate or some predestined force led me across the world to experience and enjoy their never-failing delight.

'I never saw the homestead look more beautiful than it did from the white rock when I was waiting to start my run. The fog which had spoilt much of the morning gave way to the breathless stillness of an autumn evening, and the low sun caught the poplar trees and made them look like pillars of frozen fire around some heathen shrine. Red roofs and the sordidness of corrugated iron all became attractive at this range and even the dozens of cars made a delightful pattern in miniature. The long slope of the homestead paddocks carried my eyes down to the deep, still pool of the Grasmere Lake, the water blackened by the inky green of the bush on the far, steep shore; and all round the valley stood the great ramparts which have become as familiar to me as the fingers on my hand.'

It is my earnest hope that, despite all changes in human habits or through the discoveries of science, another and another generation will follow the paths that I have trodden and continue to use this magnificent country for the pastoral purpose that man and his domestic animals have pursued for thousands of years. Besides living and loving that rather primitive occupation I have devoted most of my spare time and energy to ensuring that in an ever-changing world it may survive and continue to give spiritual satisfaction to those who undertake its challenge.

If I have succeeded in some small measure I shall be content. If I, and those with whom I have worked, have failed, then a very fine and inspiring chapter in New Zealand history will be closed and only those who care to reopen the book and read the story will understand its inspiration.